Reptiles: A Very Short Introduction

T0055144

Very Short Introductions available now:

For more information visit our website

www.oup.com/vsi/

T. S. Kemp

REPTILES

A Very Short Introduction

OXFORD
UNIVERSITY PRESS

OXFORD
UNIVERSITY PRESS

Great Clarendon Street, Oxford, OX2 6DP,
United Kingdom

Oxford University Press is a department of the University of Oxford.
It furthers the University's objective of excellence in research, scholarship,
and education by publishing worldwide. Oxford is a registered trade mark of
Oxford University Press in the UK and in certain other countries

© T. S. Kemp 2019

The moral rights of the author have been asserted

First edition published in 2019

Impression: 1

Published in the United States of America by Oxford University Press
198 Madison Avenue, New York, NY 10016, United States of America

British Library Cataloguing in Publication Data
Data available

Library of Congress Control Number: 2018957113

ISBN 978-0-19-880641-7

Printed in Great Britain by
Ashford Colour Press Ltd, Gosport, Hampshire

Contents

List of illustrations

Chapter 1
What is a reptile?

There are many popular misconceptions about reptiles, and none more so than those found in a quote of 1797 attributed (possibly wrongly) to the father of animal classification, Carl Linnaeus: 'Reptiles are abhorrent because of their cold body, pale color, cartilaginous skeleton, filthy skin, fierce aspect, calculating eye, offensive smell, harsh voice, squalid habitation, and terrible venom; wherefore their Creator has not exerted his powers to make many of them.' In fact, as we shall see, practically every one of these claims is untrue. Reptiles are frequently warm to the touch, often colourful, have hard bony skeletons, skin as clean as any animal, and no particular smell. Most are silent and few of them are venomous. Furthermore, evolution has generated a considerable number of them—there are close to 10,000 different species, which is similar to the number of birds and getting on for twice the number of mammals.

Today five different kinds of living animals make up the Class Reptilia. These are the chelonians (turtles and tortoises); the lizards; the snakes; the crocodilians; and the single rather lizard-like species of *Sphenodon*, the tuatara of New Zealand. Despite their huge range of body forms and ways of life, from a completely limbless, venomous snake that swallows whole prey larger than its head to flipper-bearing, fish- and seagrass-eating marine turtles, zoologists recognize them all as reptiles because of a range of fundamental characters that they all share. For example, the skin is dry and scaly, the two kidneys are compact, the urine waste can be excreted as a solid, and the heart has a partial internal division. Most importantly, reptile eggs are fertilized inside the female's body by sperm introduced via a male penis. Food and water reserves are added to them, and a protective outer shell or leathery membrane is laid down around them. This kind of egg is called *amniotic* or *cleidoic*, and is laid on dry land rather than in water. Inside it, the embryo safely develops until it hatches as a miniature adult.

The reptilian skeleton also has numerous unique characteristics and these are preserved in fossils. From their fossil record, we know that reptiles have a very long history stretching back around 320 million years, and that the five living Orders represent only a few of the many and varied kinds that evolved. From about 200 to about 66 million years ago, reptiles completely dominated the Earth's large animal fauna: dinosaurs on dry land, pterosaurs in the air, and ichthyosaurs, plesiosaurs, and others in the seas.

We also know from fossils that the evolutionary branch eventually giving rise to the mammals separated from the other reptiles almost at the very start of this reptilian history. Furthermore, the birds evolved from one of the dinosaur groups. This actually creates a slight problem in the naming of the groups. Modern taxonomists insist that properly named groups of organisms must consist of *all* the species that descended from a single ancestor. This gets rid of any confusion about how different groups are related to one another in the evolutionary tree of life. However, neither mammals nor birds are considered to *be* reptiles but to have evolved *from* them. Therefore, strictly speaking the group 'Reptilia' should not be used for what is left after the birds and mammals are removed. To solve this (Figure 1), we formally classify reptiles, birds, and mammals together as a large group called Amniota (referring to the nature of the egg). The mammals

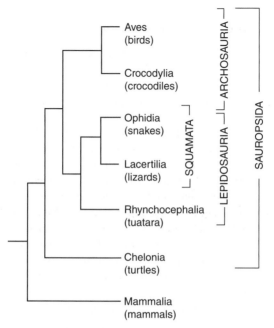

1. The relationships of the living groups of reptiles.

and their more 'reptile-like' fossil predecessors are included
together in one subgroup called Synapsida. The reptiles plus
their bird descendants are placed in a second subgroup
called Sauropsida. (Sometimes the reptiles are informally
and rather clumsily referred to as 'non-avian, non-mammalian
amniotes'.)

However, this rigorously strict correctness of terminology need
not detract from the fact that there is a perfectly good biological
sense in which reptiles are a distinctive kind, or grade, of animals.
Their lifestyles, and the roles they play in their communities,
are well understood and described by the term 'reptilian', as we
shall see in the rest of this chapter.

A brief who's who of living reptiles

Almost 10,000 living species of the Class Reptilia have been described, and no doubt a good many more are yet to be discovered. This compares with about 5,500 mammals, 8,000 amphibians, and 10,000 birds, and means that reptiles have a similar significance for the world's biological communities as do these other land vertebrate Classes. The number of species within the five individual Orders making up the Class Reptilia is, however, much less equitable. Over half of them are lizards, Order Lacertilia, while snakes, Order Ophidia (Serpentes), have around 3,700 members. The Order Chelonia (Testudines) accounts for a modest 340 species, and the Crocodylia a mere twenty-five. The final Order, Rhynchocephalia, is represented by the single species of tuatara. No one fully understands why there are such variable numbers of the different reptile groups. In part perhaps, it is simply that there are a lot more niches available for small, land-living animals like lizards and snakes than for larger-bodied semi-aquatic ones like crocodiles.

Looking at the evolutionary relationships amongst these five Orders of living reptiles (Figure 1), we find that the lizards and snakes are descended from a single common ancestor, and together they make up the Squamata. The Rhynchocephalia are the closest relatives of the squamates, technically called their sister group, and the two together complete a group known as the Lepidosauria. The crocodylians are the living members of the other main reptilian group, the Archosauria (which technically also includes the birds, as we have seen). The evolutionary position of the chelonians within this framework is not certain. The trouble is that chelonians are so different from all other reptiles that there are no characters pointing clearly to a relationship between them and any of the other reptile groups. They consist almost entirely of a mixture of features that are either found in more or less all reptiles or that are unique specializations of their own. None of these helps decide their relationships. As a result, different taxonomists have suggested almost every possible sister-group relationship: to all the other reptiles combined, to the archosaurs as a whole or to the crocodylians in particular, to the squamates, and even to the mammals. Unlike many other groups of organisms, evidence from the DNA sequences of their genes has not so far been able to resolve the question to everybody's satisfaction, although the most recent evidence does tend to support a relationship with the archosaurs.

Life on dry land

Four hundred million years ago, before terrestrial vertebrates had evolved, the idea of a vertebrate animal living permanently on dry land would have seemed absurd. Losing water from the body by evaporation would have proved fatal, even if the extreme changes in temperature that occur without the buffering effect of water had not killed it by overheating or freezing first. Without the buoyancy effect of being submerged in water, the creature would have been incapable of movement beyond the poorly controlled wriggling and thrashing of the body that is the best a stranded fish can manage. And if even these hazards were overcome, there would still be an utter dependency on water for transferring the male sperm to fertilize the eggs, and for the young to develop. The change from extracting oxygen dissolved in water to breathing in air requires a quite different kind of organ, a lung rather than gills. Even changing from water to air as the medium for receiving sensory information about the environment would have presented problems.

Yet terrestrial vertebrates did evolve step by step from fish, until by about 320 million years ago there were several different kinds. Most were only partially terrestrial, and continued to need water for protection during much of their life, and for reproduction. But one new lineage successfully solved all the problems, and in due course underwent the great evolutionary radiation into all the different kinds of reptiles. Even those that later returned in part or completely to an aquatic life, ranging from the extinct marine reptiles of the Mesozoic to the sea snakes, water monitors, and sea turtles of today, took their new kind of biology with them. The nature of reptiles can therefore best be understood in terms of their adaptations to fully terrestrial life.

The skin (Figure 2) is a reptile's first line of defence against excessive water loss, and also against physical hazards posed by the environment. It is made up of two main parts. The outer part is called the *epidermis*, a layer of living cells that continuously divides. Each new cell produces a hard protein called keratin which eventually fills it up, and the cell dies. But it stays on the surface, and therefore a relatively thick layer of keratin is created that makes a tough protective and waterproof layer. The other part of the skin is called the *dermis*. It lies below the epidermis and is made up of soft connective tissue in which the blood vessels supplying the living cells are embedded. There are also sensory nerves for detecting various kinds of information about the immediate surroundings. Other cells in the dermis are responsible for producing lipid molecules, which spread right through the skin and give it its waterproof properties.

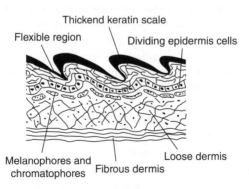

2. **The structure of the skin of a lizard.**

Although the basic structure of the skin is the same in all reptiles, it differs in detail from species to species depending on their preferred habitat. For example, the skin of a desert-dwelling lizard or snake has much more lipid and so is almost completely waterproof compared to that found in habitually aquatic species such as crocodilians or freshwater turtles. The way in which the skin is formed into scales also differs in different kinds of reptiles. Scales are thick regions of the epidermal keratin layer that are separated from one another by thin folded regions (Figure 2). This arrangement provides the animal with the physical protection of the layer of hard keratin while still allowing enough flexibility of the body for movement.

In addition to the keratin scales, some reptiles such as skinks and crocodiles also have what are called *osteoderms*. These are bony scales that are laid down within the dermal layer to provide extra physical protection. The most extreme development of osteoderms is found in the chelonians. The tortoise or turtle shell that we see is actually an interlocking pattern of very large epidermal scales called scutes. Beneath the scutes there is a layer of large, interlocking osteoderms. The arrangement of the osteoderms is different from the arrangement of the scutes, which increases the strength and rigidity of the shell.

While thinking about the skin, another of its functions not directly related to water conservation is coloration. Many snakes and lizards have vivid colour patterns for various purposes. The bold black, yellow, and red banding of the venomous coral snakes warns predators not to attack them. Certain harmless snakes, called mimics, have taken advantage of this and evolved a banding similar to a coral snake, and so gain protection for themselves. Other species, such as the green mamba, use their colour as camouflage against bright foliage, or for social and sexual signalling to members of the same species, as in the aggression and mating displays of many male agamids and chameleons. The colours themselves are due to a combination of cells called *melanophores* which contain black pigment, and *chromatophores* that contain yellow, orange, or red pigments. Beneath the pigment cells there may be a layer of cells called *iridophores* that contain reflective crystals and are responsible for the brilliant iridescence of many. The legendary ability of chameleons to change their colour pattern so rapidly is due to two processes. The melanin pigment granules can either disperse throughout the melanophore cells making the skin appear darker, or concentrate into a small space making it lighter. Added to this, the exact spacing of the crystals in the iridophores can alter very, very slightly, which changes the particular wavelengths of light they reflect, and therefore which colour they appear.

After evaporation from the body surface, the next largest cause of potential water loss in land animals is the need to remove excess nitrogen from the protein contained in their food by excreting urine. Reptiles such as terrapins and crocodiles have ample access to fresh water and mostly excrete the water-soluble, nitrogen-containing molecules, ammonia and urea. But species for which water availability is critical, especially lizards and snakes living in arid regions, mainly use an excretory product called uric acid. This is more wasteful of energy because each molecule of uric acid contains five carbon atoms compared to the single one in a urea molecule. But uric acid has the great advantage of being almost insoluble in water. This means that it is easy to reabsorb the water from the urine, precipitate out the uric acid as a semi-solid paste, and expel it with hardly any water loss at all. There is a chamber on the underside of the body just in front of the tail, called the *cloaca*. The cells of the cloacal wall have the ability to reabsorb water, and it is here that the urine is actually dried. The reptile anus also opens into the cloaca rather than directly to the outside, and so water from the animal's faeces can be reabsorbed before they are deposited, which is also an important way of conserving water.

Marine reptiles, notably the sea turtles, have a special problem of water conservation. Their body fluids are only about half as concentrated as the surrounding sea water, so they are always tending to lose water by osmosis out, and to gain excess salt from their food. To combat this, they have evolved specialized salt-secreting glands that produce concentrated sodium chloride. These open into the eye socket and produce salty tears. The marine iguanas of the Galapagos Islands have the same problem, and they have salt glands opening into their noses, from where they 'sneeze' the salt out as a spray from time to time.

Warmth for free

Temperature is very important in the life of all organisms because almost all the physical processes and chemical reactions of the body are sensitive to it. This is particularly true for a vertebrate, with its large brain and the complex behaviour it controls, and there is a fairly narrow range of body temperatures over which the animal can be fully active. A well-known study thirty years ago on the wandering garter snake illustrates this perfectly (Figure 3). Several of its normal daily activities such as crawling, swimming, tongue flicking, and digestion were at a maximum when the body temperature was between 30 and 35°C. Activity fell away as the body temperature decreased, and at about 10°C had more or less ceased. Similarly, as the temperature rose the levels of these activities also fell, and at about 40°C had ceased. The optimum body temperature of reptiles differs from species to species, depending largely on its habitat. In desert-living iguanids it is as high as 40°C, while in contrast the tuatara lives in the cool climate of New Zealand and is most active at night. Its optimum body temperature is around 18°C. As with the garter snake in the laboratory, reptiles in the wild become increasingly inactive, and eventually completely torpid, as their body temperature falls. Increasing the body temperature also reduces activity, and in this case can soon be fatal. The range of body

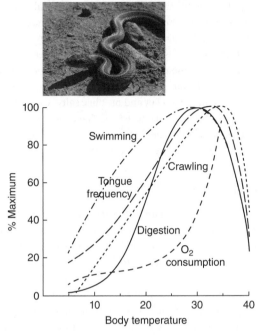

3. **Activity levels and body temperature in the garter snake.**

temperatures over which the reptile can lead its normal, active life is invariably narrower than the range of external temperatures it faces on both a daily and a seasonal basis.

These observations show how important it is in the life of a reptile, to be able to regulate the body temperature as close as possible to the optimal value. Studies of reptiles in their natural habitat highlight just how remarkably effective they can be at regulating body temperature, in the face of a wide range of environmental conditions. The way in which this is done is called *ectothermy*, because virtually all the heat of the body is acquired from the environment. This contrasts with *endothermy*, the way mammals and birds regulate body temperature, where heat is

10

generated by the chemical reactions in their body tissues. Endothermy is a more reliable source of heat, but vastly more costly in terms of how much food has to be consumed and oxygen breathed in to produce enough. The metabolic rate and therefore the daily food requirement of a resting reptile is only about one-tenth that of a similar sized mammal. Ultimately, the source of body heat of a reptile is the sun. This may be acquired directly by basking, or indirectly by absorbing heat from the ground or rocks that have already been warmed up. To keep the body temperature close enough to the optimum, the amount of heat gained by the body has to be continually balanced against the amount of heat it loses, and reptiles have a number of ways of controlling both these rates. Increasing the amount of heat is most easily done by basking in the sun to warm up, and this is very common behaviour in lizards, snakes, and crocodiles in the early morning. Typically, the animal starts its day lying sideways on to the sun to expose as large an area of the skin as possible. Later in the day, as the body temperature rises, they may turn their body around to face the sun and expose a smaller area, therefore reducing the heat input. Once the right temperature has been reached, reptiles seek cooler, shadier places in crevices and amongst vegetation, or in the case of crocodiles and water monitors, by entering the water. Desert-dwelling lizards are frequently burrowers, which is a simple way to avoid the full heat of the day.

On top of this kind of behaviour, various species have more specialized physiological ways of keeping the body temperature within their activity range. One is the ability to alter how rapidly the heart beats. A higher heart rate increases the rate of flow of blood through the skin and so more heat is transferred between the body and the environment, either inwards or outwards depending on the outside temperature. All reptiles can do this to some extent, but the most spectacular example is the marine iguana of the Galapagos Islands. This lizard feeds on seaweed by diving into the cold sea, and as it does so the heart rate falls to

2–3 beats per minute greatly reducing heat loss to the water. When it emerges, the heart rate recovers as it lies in the full sun and it rapidly warms up again. Both sea snakes and marine turtles similarly decrease their heart rate when diving compared to when they lie floating at the water's surface. Another physiological way of controlling body temperature that is found in many lizards such as the desert iguana is to vary the darkness of the skin by altering the distribution of the melanin particles in the melaophore cells. Spreading them makes the skin darker so that it absorbs the sun's heat faster, and vice versa.

Reproducing out of water

The *cleidoic* ('hidden'), or *amniote* ('with amniotic membrane') egg is the single most important adaptation of reptiles, because it can be laid and develop on dry land, allowing animals that possess one to live a completely terrestrial existence. Until it evolved, virtually all vertebrates had needed a body of water in which to shed their gametes, so that the sperm could swim to and fertilize the eggs, and in which the fertilized eggs could develop before hatching into aquatic larvae. A very small number of fish and amphibian species also fertilize the eggs by the male inserting sperm directly into the female's oviduct, where she retains the eggs before giving birth to live young. However, the young still require water. A few North American salamanders do go so far as to lay their eggs on land, although even they need permanently cool and humid conditions to develop and hatch as miniature adults.

We have already met the reptilian cloaca, the recess on the underside of the body into which both the urinary ducts and the rectum open, and which can reabsorb water from the urine and faeces, and temporarily store these waste products. The cloaca has a third important function, because the female reproductive ducts also open from it. In the male, an extension of the wall of his cloaca forms a penis, which is used during copulation to introduce sperm into the female's cloaca. The sperm swim forwards from here into the oviducts, where they meet the eggs and fertilize them.

Once this has happened, the cells lining the wall of the oviduct lay down a soft, tough membrane around each egg. In many turtles, the crocodiles, and also in the past the dinosaurs, crystals of calcium carbonate are added to the membrane which harden it into a shell. Forming this shell around the egg is remarkable enough, but what happens inside it is even more extraordinary, and we have to understand how the egg develops to appreciate how all the necessities of life of an embryo are provided within it. Three thin, bag-like membranes grow around the embryo (Figure 4).

The first one is called the *amniotic fold*. It grows out of part of the embryo, enlarges, and spreads until it completely surrounds the whole embryo. The outer wall of this bag is called the *chorion* and it lies tightly against the inside of the egg shell. The inner wall in called the *amnion* and this loosely surrounds the embryo, creating a fluid-filled enclosure called the amniotic cavity. Meanwhile two more membranous bags develop, this time growing out of the developing gut of the embryo. One is called the *yolk sac* and is filled with yolk granules, which are concentrated food particles needed to sustain the embryo all the way to hatching. The second sac is called the *allantois* (named after the Greek word for sausage-shaped), which grows until it fills much of the space between the chorion and the amnion. The allantois has two purposes. It is where the developing embryo deposits its excretory products, and also, by developing a network of fine blood vessels near the surface of the egg, it acts as a lung. These blood vessels carry the gases, oxygen in and carbon dioxide out, between the embryo and the environment.

Together therefore, the different parts of the cleidoic egg provide all the embryo needs to complete its development without having to be laid in water, or hatch as an aquatic larva. The shell, whether a soft leathery one or a hard calcified one, offers physical

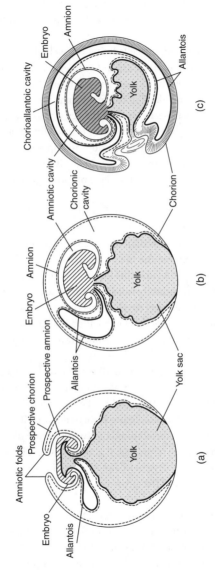

4. a–c, Three successive stages in the development of the cleidoic egg.

protection and is permeable to gases. It also greatly reduces loss of water, and in most reptiles it can even absorb a certain amount of water from a humid atmosphere. The amnion and the amniotic cavity it bounds create a shock-absorbing, fluid-filled haven, the yolk sac provides the nourishment, and the allantois is responsible for breathing, and for waste storage. The amniotic egg has been well described as providing a private little pond for the embryo.

This ability of the cleidoic egg to develop on land does have a downside: it cannot develop in water at all. In effect the embryo would drown because, unlike amphibian tadpoles and the larvae of most fish species, it has no water-breathing gills. This means that the groups of reptiles that have reinvaded the aquatic habitat must do one of two things. Some return to land to lay their eggs, such as the crocodiles which lay them in nests dug in the soil near the river bank, and the sea turtles, which bury them in a nest on a sandy beach. Others, notably the sea snakes, keep the eggs inside the mother's body, where they develop until they reach a large enough size to be able to hunt for themselves. This ability to bear the young live is called *viviparity*, and from the evidence of a number of beautifully preserved fossils that still have tiny juveniles within the body cavity, we know that the Mesozoic ichthyosaurs described in Chapter 2 did the same.

A few lizards and many snakes are also viviparous, but for an entirely different reason. The fertilized eggs remain in the mother's body, and so the developing embryos can take advantage of her basking behaviour to keep their own temperature higher. Several of these species live in relatively cool regions of the world, where viviparity is an advantage for reproducing more quickly. The European adder, for example, is the only snake that regularly lives above the Arctic Circle. The female produces about a dozen live offspring every other summer, and juveniles are born with their venom already functional. However, many temperate species are still egg layers, and several snakes, such as the boas, living in the warmer regions of Africa are viviparous, so there is no simple correlation with habitat. In fact viviparity is a good illustration of how evolutionary change is always a compromise, or trade-off, between costs and benefits. It certainly increases how fast the young grow, but carrying the embryos inside her body increases the amount of food the mother must collect, and therefore the extent to which she is exposed to predators as she forages.

Getting around

For their life on land, reptiles needed to evolve physiological adaptations to prevent too much loss of water, and to regulate the body temperature within fairly narrow limits when they are active. They also needed to overcome a third great problem of terrestrial existence, the force of gravity. Animals submerged in water weigh virtually nothing, thanks to the buoyancy provided by the water. But the moment a typical fish, for example, finds itself stranded in air, its weight is instantly huge and it is unable to move very much at all other than by a directionless flapping of its body. For effective locomotion on land, a strong skeleton is needed to support the weight of the animal's body, and moveable limbs that can propel it in the direction it wants to go by pushing forcefully against the ground (Figure 5). This new anatomy is needed even in vertebrates that spend only short periods out of water. In fact the weight problem was solved before the problems of water loss and temperature fluctuation. In the earliest fossil land vertebrates, which never became fully land-living, we already see the four stout legs necessary for terrestrial locomotion. However, it was the reptiles that perfected this support system, and watching the speed and agility of a modern lizard running away from a predator, or a crocodile lunging after a bird, demonstrates how effective is the basic mode of reptilian locomotion.

The vertebrae that make up the length of the spinal column interlock with one another by special extensions, front and back, called *zygapophyses*. These give the whole column the strength of a single girder to carry the weight of the animal's body, a strength

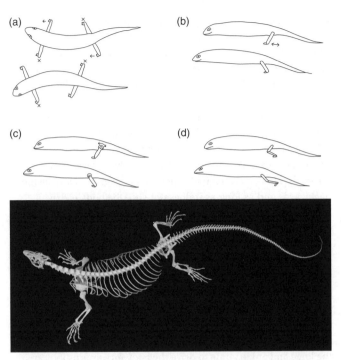

5. Locomotion in a typical reptile: (a) lateral undulation; (b) limb retraction; (c) long-axis rotation; (d) limb extension. Below, skeleton of a monitor lizard.

made even greater by the pairs of ribs partly surrounding the body that each vertebra carries. However, the zygapophyses also allow the vertebral column to be flexible, and the body can undulate from side to side like an eel, by sideways bending between vertebrae. The large shoulder and hip girdles are where the limbs connect to the trunk, and they transmit the forces that the limbs apply to the ground to move the whole body. The limbs are remarkable structures. In mechanical terms they act simultaneously like levers and extensible struts. Each leg, front and back, is made up of three parts. There is a single, large bone that fits into the socket of the limb girdle, and sticks out sideways

17

from the body. The middle part is made up of two bones attached to the first and which are held vertically. The lower part is the foot, which includes the wrist or ankle joint, and there are five toes of different lengths. The paired feet, front and back, are held wide apart and placed flat on the ground when the animal is standing.

Locomotion in a typical lizard or a crocodile is made up of several parts. The first one is lateral movements of the whole trunk (Figure 5a). Waves of sideways bending pass down the body, and as they go they make the legs swing passively backwards and forwards. However, each foot is placed on the ground as it swings backwards and as long as it does not slip, the body is driven forwards. It is then lifted up for the forwards swing ready for the next stride. The second component is called limb retraction (Figure 5b), and consists of the limb muscles actively moving the limb forwards, lowering the foot onto the ground, and then pulling it backwards. Again, as long as the foot does not slip, the body is driven forwards. Thirdly (Figure 5c), the upper limb bone rotates about its long axis, making the lower bones and foot swing relatively backwards. The fourth component (Figure 5d) occurs once the foot is behind the limb girdle, and consists of extending the knee and ankle joints to drive the body forwards as if by an extensible strut. These four components combine to make a longer overall stride, which leads to faster locomotion.

This basic reptilian way of walking and running has been modified in a number of ways in other reptiles for leading different ways of life, as we shall see in more detail later. Several lizards and the snakes have lost their legs and rely on the lateral undulation of the body alone. This works well in burrowing animals, and in the case of many snakes is a good way of living in trees, or swimming in the sea. The chelonians have done the opposite and their rigid shell prevents any undulation of the body at all. The land tortoises rely on the movement of the limbs alone, while in the marine turtles the limbs have been modified into extremely efficient swimming paddles, but which are very inefficient on land.

Chapter 2
History of reptiles

In 1811 Joseph and Mary Anning, the children of an impoverished widow living in the southern English coastal village of Lyme Regis, found on the beach the fossilized skeleton of a huge creature with a vaguely crocodile-like head but the like of which had never been seen before. At about the same time a young doctor and amateur geologist from Lewes in Sussex called Gideon Mantell was coming across large bones and strange teeth of some huge animal that seemed to be land living and plant eating. How reptiles were viewed by scientists and the public alike was never to be the same again, for the Annings' fossil was that of a giant sea reptile, soon to be named ichthyosaurs, and before long Mary had grown famous for her many discoveries of these and the equally bizarre reptilian sea monsters called plesiosaurs. As for Mantell's bones, they belonged to what soon became known as dinosaurs, or 'terrible lizards'. (In fact, the jaw of a meat-eating dinosaur called *Megalosaurus* was already being displayed in the University Museum of Oxford University, although its significance had not been fully realized.) Over the course of the rest of the 19th century, many other discoveries from other parts of the world continued to reveal an astonishing diversity of reptiles during what is called the Mesozoic Era, the time appropriately christened 'The age of reptiles'.

Origin and early evolution of reptiles

In fact, the conquest of land by the reptiles had started much earlier than the Mesozoic. By about 410 million years ago, during the Silurian Period of the Earth's history, a few simple plants, more robust and tough than most of the soft, weedy algae of the time, had evolved the ability to live in the shallows and margins of lakes and rivers. They were resistant to drying out, and their stems were strengthened by fibres allowing them to grow up above the water surface so that their reproductive spores could be carried further afield by the wind. Once well established on land, these plants provided a potential new source of food for any animals that were also able to withstand the rigours of terrestrial life, and by fifty million years later a land fauna had evolved. There were various kinds of primitive arthropods such as insects, centipedes, and relatives of spiders, along with worms and snails.

It was perhaps inevitable that certain specialized fishes already living in the shallow waters and able to get some of their oxygen by means of a simple lung should start to feed on these invertebrates. At first, no doubt, they took individuals that had accidentally fallen into the water, but gradually they evolved adaptations that let them emerge onto land for brief periods and actively hunt their prey. Encouraged by increasingly abundant oxygen in the air during the Early Carboniferous Period (Figure 6), they had evolved into several different kinds. Most still needed to reach water, certainly for reproduction, because we occasionally find fossilized larval stages that have gills. Also, the water still protected them from the potentially fatal hazards of desiccation and overheating in the sun. The amphibians of today are the frogs, salamanders, and worm-like caecilians, all descendants of some of these early forms, and they still follow the same semi-aquatic lifestyle. However, one kind of early tetrapod became much more independent of water.

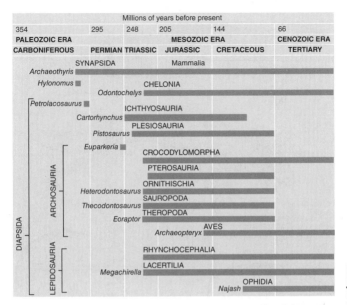

			Millions of years before present		
354	295	248	205	144	66
PALEOZOIC ERA			**MESOZOIC ERA**		**CENOZOIC ERA**
CARBONIFEROUS	**PERMIAN**	**TRIASSIC**	**JURASSIC**	**CRETACEOUS**	**TERTIARY**

SYNAPSIDA — Mammalia
Archaeothyris
Hylonomus — CHELONIA
Odontochelys
Petrolacosaurus — ICHTHYOSAURIA
Cartorhynchus — PLESIOSAURIA
Pistosurus
Euparkeria — CROCODYLOMORPHA
PTEROSAURIA
ORNITHISCHIA
Heterodontosaurus — SAUROPODA
Thecodontosaurus — THEROPODA
Eoraptor
AVES
Archaeopteryx
RHYNCHOCEPHALIA
LACERTILIA
Megachirella
OPHIDIA
Najash

DIAPSIDA / ARCHOSAURIA / LEPIDOSAURIA

History of reptiles

6. The geological timescale showing the occurrence of the main amniote groups.

There is a coal-mining area in Nova Scotia called Joggins, whose Upper Carboniferous rocks are about 312 million years old. The fossilized rotting stumps of giant lycopod trees, remains of a great tropical forest, are often found embedded in the coastal cliffs. Within these hollow petrified stumps the skeletons are sometimes found of small animals that lived at the time. Perhaps they had used the hollow stumps as safe places to hide, or perhaps, as some palaeontologists have suggested, the stumps filled with water and these unfortunate animals fell in and drowned. At any event, amongst the fossils is one called *Hylonomus* (Figure 7a). It did not have any sign of the *lateral line* grooves on its skull that house sense organs for detecting sound waves in water that are found in fishes and amphibians. Furthermore, no larval specimens of *Hylonomus* have ever been found bearing gills typical of an

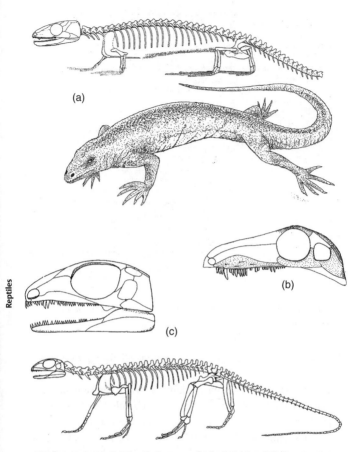

7. Early amniotes: (a) the skeleton and whole body of *Hylonomus*; (b) the skull of the synapsid *Archaeothyris*; (c) the skull and skeleton of the diapsid *Petrolacosaurus*.

aquatic larva. So while there is no direct evidence that *Hylonomus* laid cleidoic eggs on land, which would prove that it was truly an amniote, does have a number of characters of its fossilized skeleton that are found in the Amniota. We can safely assume that *Hylonomus* offers a plausible picture of what the ancestor of all the reptiles was like. It was quite small, only about 25 cm long, and the shape of the body and proportions of the legs were much like those of a typical modern lizard. A row of sharp-pointed teeth along each jaw indicates that its main food would have been small terrestrial invertebrates. One particularly important anatomical feature is that there is no gap or window in the bones of the cheek region of its skull. As we shall see in a moment, most of the later reptiles do have either one or two such windows; the skull of *Hylonomus* is therefore described as *anapsid* (without arches).

A second kind of amniote found in the tree stumps of Joggins is called *Archaeothyris* (Figure 7b) which, unlike *Hylonomus*, does have a window in its cheek, called a *temporal fenestra*. *Archaeothyris* is in fact the first member of a group called the Synapsida, the other great amniote line, and the one which eventually led to the mammals. The divergent evolutionary pathways leading on the one hand to the reptiles and birds, and on other to the mammals, are indeed ancient, for they separated almost at the very start of amniote evolution 320 million years ago.

For the next important step in our story of the reptiles, we now have to go to some shaly rocks in Kansas, where a small fossil skeleton was discovered that is similar in size and shape to *Hylonomus*, but that has two temporal fenestrae in the cheek region of its skull. Named *Petrolacosaurus* (Figure 7c), it is the earliest member of the group called the Diapsida. And it could hardly be of greater importance in reptile evolution, because the great majority of reptiles from this time onwards were also diapsids. For a long time it was actually the synapsids that dominated the land, diverging into numerous kinds of large and small carnivores and herbivores. However, everything changed 250 million years ago when one of the most dramatic events in the entire history of life took place. A sudden, worldwide extinction removed over 90 per cent of the world's species of animals and plants, in water and on land. Such mass extinctions, as they are

called, have happened from time to time throughout the history
of life on Earth, and this particular one is called the end-Permian
mass extinction. Its exact cause is not easy to pinpoint. It seems
to have been triggered by a massive volcanic eruption that spewed
out huge amounts of noxious gases into the atmosphere, killing
plants and raising the global temperature for a few million years.
The effect was a great change in the fortunes of reptiles.
The synapsids never fully recovered from their losses, but one
of the groups of diapsids, called the archosaurs, that had been
very rare beforehand began to expand and diversify.

The archosaurs: dinosaurs

Fossils of a diapsid reptile about half a metre long called
Euparkeria (Figure 8a) are occasionally found in South Africa, in
rocks about 245 million years old. Its most notable differences
from *Petrolacosaurus* are sharp, meat-eating teeth set firmly into
sockets, a space in the skull in front of the eye socket, and back
legs that are a lot longer than the front legs. These and other
features are characteristic of the Archosauria, and tell us that
Euparkeria was an agile little hunter, perhaps even able to run
for short periods on its back legs alone, a mode of locomotion
called *bipedality*. From this time on, archosaurs evolved into a
great variety of different kinds.

The dinosaurs are the most famous, as well as the most
abundant branch of the archosaurs, and about ten million years
after *Euparkeria* lived, we start to find the earliest fossils of the
three main groups. *Eoraptor* from South America was a small,
one-metre-long predator that had evolved to run permanently on
its back legs. It represents the start of the great carnivorous
branch of dinosaurs called the Theropoda, which over the course
of the remaining 135 million years of the Mesozoic evolved into
many different species. The most familiar are the tyrannosaurids,
culminating in the 12 m long, 14 tonne *Tyrannosaurus rex*
(Figure 8d), standing almost 4 m high on its massive hind limbs.

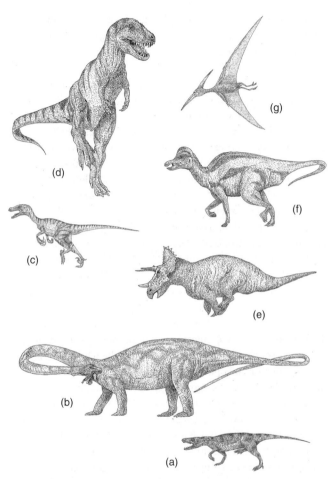

8. Mesozoic archosaurs: (a) *Euparkeria*; (b) *Diplodocus*; (c) *Velociraptor*; (d) *Tyrannosaurus rex*; (e) *Triceratops*; (f) *Corythyrosaurus*; (g) *Pteranodon*.

Its jaws carried a fearsome row of long, serrated teeth each up to 15 cm long. Other more modest sized, but equally rapacious theropods evolved. *Velociraptor* (Figure 8c), for example, was only 1.5–2 m long and would have stood no higher than a human's waist. Other species were probably scavengers rather than hunters, and at least one may have been an egg eater. Others even seem to have been herbivores.

It is now quite certain that the birds evolved from theropod dinosaurs, indeed that they should strictly be classified in the Theropoda themselves. A number of forerunners of the birds have been discovered over the last twenty years, especially in Chinese Mesozoic rocks. The first to be described was named *Sinosauropteryx*. It astonished the world, because its skeleton was covered in the impressions of feathers, until then believed to be found only in birds. However the front legs of *Sinosauropteryx* were short and not at all like wings. Since then, several other small, feathered theropod dinosaurs have come to light, strengthening even more the link to birds. *Microraptor*, for example, had long, feathered front and hind limbs and is a possible four-winged intermediate stage in the evolution of flying. The distinction between small feathered dinosaurs and birds is now quite blurred.

In 1834, bones of a dinosaur named *Thecodontosaurus* were discovered in fossilized limestone caves of south-west England, and it was soon realized from the nature of the teeth that they were the remains of a plant-eating dinosaur. It is the earliest of the second great dinosaur group, called the sauropodomorphs, which were the commonest dinosaurs of the Late Triassic world. Some of them evolved to a considerable size, such as the 10 m long *Riojasaurus* that lived in what is now South America. Although it had well-developed front legs, it could stand on its back legs, and with the help of a long neck browse on higher vegetation. A subgroup of the sauropodomorphs evolved into seriously gigantic animals, the sauropods, which are often erroneously called 'brontosaurs' (Figure 8b). They first arrived on the scene in the Early Jurassic Period and soon replaced their smaller forebears. Sauropods had four stout, columnar legs to support the huge body, a long neck to reach the vast amounts of vegetation they needed, using rows of flattened teeth with serrated edges to collect it. All weighed many tonnes, such as the 25 tonne *Brachiosaurus* of North America, but the very largest of all was *Argentinosaurus*, named from where it was found. It was around 35 m long from nose to tail, and weighed as much as 75–90 tonnes. Compare this, the greatest land animal ever to have lived, with the record weight for a full-sized modern African bull elephant of a mere 12 tonnes.

The Ornithischia make up the third major group of dinosaurs to have evolved by the end of the Triassic Period, although they were still very rare at this time. *Heterodontosaurus* from South Africa was the size of a largish dog. When it chose to, it could walk bipedally, although it probably spent most of the time on all four legs. Its teeth and jaws show us that it was essentially a herbivore that was capable of chewing its food. Although slow to start, by the middle of the Jurassic period ornithischians had become the main dinosaurian herbivores, and had evolved into several kinds, most of them very familiar to us. The ornithopods include *Iguanodon*, the dinosaur first discovered by Mantell and one of the first to be appreciated as a giant reptile. Like all ornithischians, *Iguanodon* had a beak rather than teeth at the front of its mouth for cropping vegetation, and rows of serrated, leaf-like teeth along the jaws for chewing it up. The duck-billed dinosaurs, such as *Corthyrosaurus* (Figure 8f), were the commonest ornithischians. They had had broad heads, and hundreds of closely packed teeth for dealing with tough food such as pine needles. Some of them also had strange hollow crests on their head, which may have been for recognizing one another and for amplifying the sounds the animals made in communicating with one another. Other familiar groups were the stegosaurs with huge bony plates running down the back, perhaps for social signalling or helping to keep the body cool. The ankylosaurs were protected by heavy armour-plating, and the ceratopsians like *Triceratops* (Figure 8e) had a huge protective neck-frill of bone, and were armed with up to three horns on the head to fight off predators or tussle with rivals.

For 135 million years, until their extinction at the end of the Cretaceous Period, the dinosaurs were far and away the most abundant and diverse large-bodied terrestrial vertebrates. Certainly they shared their world with other kinds of vertebrates, including many mammals, early lizards, and others, that were all far smaller animals. There has been a great deal of discussion amongst palaeontologists about why dinosaurs were so successful, and whether they were perhaps very different from today's reptiles. For many years, the mere fact that they were classified as reptiles led people to assume they were ectothermic or 'cold-blooded', and relatively slow and inactive, as was the view then about modern reptiles. The more careful modern ideas about classification and about modern reptile biology have discredited such a simple opinion. The endothermic birds are technically dinosaurs, so why should other dinosaurs not have had bird-like temperature physiology? We need to judge the nature of dinosaur biology on the evidence of the actual fossils, and furthermore it would be wrong to think that all dinosaurs were biologically just the same as one another. The large-bodied kinds such as sauropods, tyrannosaurids, and most of the ornithischians could probably have kept their body temperature at a more or less constant level simply because the body surface area through which heat was lost was small compared to the mass of the body that was generating the heat in the first place. They would have lost heat from the surface slowly enough that even at night the body temperature fell scarcely at all. Indeed, the need to lose rather than retain heat may have been a greater problem for them, particularly as the Mesozoic Era was a time of generally much warmer climates than today. Air-filled cavities in the vertebrae and other bones of the sauropods probably helped in this, as did the back plates of the stegosaurs and the neck-frill of the ceratopsians. At the other extreme, the feather-like structures in the skin of small-bodied theropod dinosaurs, both small species and juveniles of larger ones, must have provided insulation to reduce heat loss. This only makes sense if they were endotherms using their own body heat to keep at the right temperature.

Another aspect of dinosaur biology that reminds us of birds comes from the discovery of fossil sites that consist of large numbers of nests with eggs in them. Communal nests were dug in the ground by a group of adults, who probably stayed in the vicinity to protect the eggs and the hatchlings. It suggests that dinosaurs had a complex level of parental behaviour similar to that seen in crocodiles today.

The archosaurs: pterosaurs

While the dinosaurs were roaming the surface of the land, another group of archosaurs had taken to the air and evolved true flight. The pterosaurs (Figure 8g) had a hugely elongated fourth finger that can be seen in some extremely well-preserved specimens to have supported a leathery and filament covered membrane, rather like a bat's wing. Their skeletons were very lightly built and in some fossils the body is enveloped by fine, hair-like filaments, showing us that they were insulated against heat loss, proof enough that they were endotherms with the high metabolic rate that is needed for active flight. Pterosaurs came in many sizes, from as small as a sparrow to flying giants with a wingspan of several metres that were able to launch themselves from their clifftop roosts and glide over the coastal waters. They used long, very sharp-pointed teeth or a narrow, pointed beak to catch fish swimming near the surface. *Pterodaustro* was the oddest of the pterosaurs. It had as many as 1,000 extremely fine and closely packed bristle-like teeth on its lower jaw. The jaw itself was curved upwards and acted like a scoop for filtering microscopic plankton from the water as the animal flew along just above the surface. The largest pterosaur ever found is called *Quetzelcoatlus* and comes from Texas. It had a wingspan of 11 m and weighed something like 200 kg, making it by far the largest flying animal ever to have lived. It is unusual in having no teeth at all, but there was probably a lightweight horny beak instead. *Quetzelcoatlus* lived a long way inland from the coast, and it seems likely that it made its living as a scavenger of corpses of dead dinosaurs, like a huge vulture.

The archosaurs: crocodiles

Crocodiles were the third main group of Mesozoic archosaurs. Compared to the dinosaurs there were a good deal fewer of them, although there were many more species than are alive today: well over a hundred species have been found. Furthermore, during the Mesozoic crocodiles had adapted to a greater range of habitats and ways of life than the very conservative modern representatives. The first to evolve were relatively small archosaurs, only about one metre in length. Curiously they had distinctly longer back legs than front legs, because they had evolved from bipedal ancestors, and to some extent were still able to run in this fashion. Most crocodiles soon increased in size and evolved longer front legs for fast quadrupedal running. These were rapacious hunters, fully adapted to living on dry land. One, the North African *Sarcosuchus*, was a real giant amongst crocodiles with a head over one and a half metres long, a body 12 m long, and a weight of around 8 tonnes. It could easily have caught and consumed the smaller dinosaurs. At the other extreme of lifestyle, *Simosuchus* from Madagascar was a mere 0.75 m long. The short tail shows that it was adapted mainly for living on land, and the short snout and clover-leaf shaped teeth indicate that it was a herbivore, most unusually for a crocodile. Other Mesozoic crocodiles adopted the typically amphibious life of modern crocodiles. A few, such as the 7 m long *Plesiosuchus*, even evolved into permanently marine creatures that were no longer able to emerge onto land at all. The body was streamlined, the limbs were reduced to paddles, and the tail carried a fin for swimming. Such creatures must surely have been incapable of coming onto the shore, and they were probably the only archosaurs ever to have borne their young live, although we have no direct evidence of this.

Marine Mesozoic reptiles

While the surface of the Earth and its skies were dominated by the archosaurs, other groups of reptiles had come into worldwide prominence in the Mesozoic seas. Soon after the great mass extinction of 250 million years ago that closed the Palaeozoic and ushered in the beginning of the Mesozoic Era, several new kinds of reptiles appear in the fossil record that had evolved adaptations for life in the sea. This, the Triassic Period, was also a time when many new fish species evolved and spread, and the new reptile groups quickly exploited this rich new source of food. Most, perhaps all of them, evolved from early diapsids that had existed before the split into archosaurs and lepidosaurs. Some of the new groups were quite short-lived, such as the strange placodonts (Figure 9b), named for their huge, pebble-like teeth. Their legs were short, but were still able to carry the heavily built body on land. However, their fossils are always found in rocks laid down in the sea, and therefore we believe they led a semi-marine, walrus-like life, using their teeth to crush hard-shelled prey like molluscs and crabs. The thalattosaurs (Figure 9a) were another kind of marine reptile that only existed for a short time before going extinct. They were elongated slender-bodied animals, some measuring as much as 4 m in length from nose to tail tip. They also had four slender legs, showing that, like placodonts, they could emerge onto land to lay their eggs. But mostly they swam in the sea using a long, flattened tail, and catching small fish with their slender, sharp-toothed jaws.

Other new groups of reptile that arose in the Triassic went on to become very important members of the marine habitat for the rest of the Mesozoic Era. One of these was the plesiosaurs (Figure 9c), extraordinary creatures that have been described whimsically as a snake threaded through a turtle shell. They had a very long neck made of about thirty vertebrae, and a shorter, narrow tail for steering through the water. The body in between was like a flat box made up of the vertebrae and ribs above, and bony plates

9. Mesozoic marine reptiles: (a) *Thalattosaurus*; (b) *Placodus*;
(c) the plesiosaur *Elasmosaurus*; (d) *Mosasaurus*; (e) *Ichthyosaurus*;
(f) the pliosaur *Nothosaurus*.

below. Plesiosaurs swam using four elongated, paddle-like legs
that beat up and down like the wings of a bird. The head was very
small for the size of the animal, and the jaws had long, sharply
pointed teeth for catching fish. Being at the end of such a long,

flexible neck, plesiosaurs were able to grab prey in their jaws from over a wide area, without having to swim fast to catch up with it. Pliosaurs (Figure 8f) were contemporary relatives of plesiosaurs, and the two together are classified as the Sauropterygia. The difference is that pliosaurs had shorter necks and very much bigger heads. They were altogether faster, more powerful swimmers and actively hunted larger prey, including fish no doubt, but probably other marine reptiles as well. Their way of life might best be compared to that of the orcas, the killer whales of today.

A very strange fossil was found in China in 2011 and named *Cartorhynchus*. It was small, only about 40 cm long, and had a large head and four short, stubby feet in which the main bones were still recognizable and could still carry it around clumsily on land, like a modern sea lion. It lived about 248 million years ago, which was right at the start of the Triassic, as life was just beginning to recover from the mass extinction. It is the earliest member of another important kind of Mesozoic reptile, the ichthyosaurs (Figure 9e), and is one of the rarely discovered cases of an intermediate grade fossil. The name ichthyosaur is from the Latin for 'fish-lizard', which aptly describes the Ichthyosauria, the most modified of all the marine reptiles. Fossil specimens are often extremely well preserved, and this sometimes includes an impression of the animal's skin. The body was shaped like a typical fish, flattened from side to side and streamlined from its narrow snout to its large, fish-like tail fin or fluke. There was even a shark-like dorsal fin in the middle of its back. As for the limbs, these were no more than small fin-like structures covered in tough skin and supported by fingers of simple interlocking bones. Ichthyosaurs were adapted for high-speed swimming like a modern tunny. Waves of contraction of the body from front to back created powerful side to side movement of the tail fin, while the limbs were only used for steering. They were extremely agile swimmers, and fed by chasing and snapping up fish in their narrow jaws. A few fossils still preserve the gut contents, which in addition to fish scales sometimes include the hooks of ammonites. Being completely incapable of emerging onto land, it was always believed that ichthyosaurs must have borne their young live. It is very gratifying, therefore, to find occasional specimens that contain the actual skeletons of tiny young inside the body cavity. One famous specimen even has a juvenile apparently in the act of emerging from its mother's birth canal.

Ichthyosaurs were all much the same general shape, although they did differ in size. The smallest adults were about 2 m long while the largest, called *Shonisaurus*, reached as much as 20 m.

The placodonts, thalattosaurs, plesiosaurs, and ichthyosaurs were all exclusively marine animals. During the Mesozoic, two other reptile groups that were mainly terrestrial did have some members that were fully adapted for life in the sea. One of these was the crocodiles which we have already met. Another group was called the mosasaurs (Figure 9d), which are related to the monitor lizards. They evolved to the largest size ever achieved by lizards, some reaching over 15 m in length. Like ichthyosaurs, they swam by means of a large fish-like tail fin and used their short, webbed legs for steering. Although they were giant lizards, they lived permanently in the sea and even bore their young live, again like ichthyosaurs. Indeed, the general similarity in way of life between mosasaurs and ichthyosaurs means they must have been competitors for food. This might explain why the increasing success of the mosasaurs coincided with the decline and final disappearance of the ichthyosaurs during the Late Cretaceous, some twenty million years before the end of the Mesozoic.

The end of the age of reptiles: the great extinction

Sixty-six million years ago the Mesozoic Era ended, and with it the 'age of reptiles'. It was caused by another mass extinction, like the one that had ended the Palaeozoic Era 184 million years earlier.

This one is called the end-Cretaceous (or K-T) mass extinction, and not a single species of dinosaur (not counting birds), pterosaur, sauropterygian, or mosasaur survived. There is still a good deal of argument amongst palaeontologists about the cause, because the study of the rocks laid down at the time provides evidence for two possible triggers. The most popular one is that a huge, 10 km meteor travelling at half the speed of light hit the Earth. The remains of a massive undersea crater lie just off the southern Caribbean coast of Mexico, which dates to the time of the extinction. It would have caused unimaginably huge tsunamis that spread out across the oceans, wildfires that consumed the forests, and would have filled the atmosphere with water vapour and vaporized rock lasting for months or years that was fatal to much of life.

However, there is evidence for a second possible cause. The centre of India is covered with a huge area of thick volcanic rocks, called the Deccan Traps, laid down at the time of the mass extinction. The extreme volcanic eruptions would have thrown out into the atmosphere millions of tonnes of gases, including sulphur dioxide which causes acid rain, and carbon dioxide which causes heating of the atmosphere by the greenhouse effect. Yet a third possible factor behind the extinction was a large fall in the sea level, which would have had a considerable effect on the climate of the time. It seems most likely that the mass extinction resulted from a combination of these factors. Together they had a much greater effect on the environment than any one of them alone would have done.

Discovering the cause of the extinction is made even harder when we look at how it actually affected the animals, for there were also a good many survivors as well as all the losers. While some of the main reptile groups, especially the dominant ones, disappeared completely, others suffered a loss, even a severe loss, but enough of them survived to begin a new evolutionary radiation once suitable conditions had returned. Many lizards, snakes, and chelonians survived, helped perhaps by their smaller body size. More surprising were the crocodiles. Despite being almost as much a part of the dominant Mesozoic reptile fauna as dinosaurs, enough of them survived to carry on into the Tertiary, and to evolve into their present-day abundance of species. Similarly, numerous species of birds and mammals survived the mass extinction to give rise in due course to their own hugely successful Tertiary radiations.

The rise of the modern reptiles: lizards, snakes, and chelonians

Throughout the Mesozoic, the dinosaurs were by far the dominant land animals, although at night the world teemed with a diverse array of active little mouse- to cat-sized mammals. Compared to both these kinds of animals, we find far fewer fossils of lepidosaurs, the lizards, snakes and sphenodonts. This is partly because they were less likely to be preserved due to their small size and delicate skeletons, but probably also because they really were scarcer. The Lepidosauria had actually evolved very soon after the archosaurs. Rhynchocephalians, the reptiles still represented today by *Sphenodon*, have been discovered in 240 million year old Middle Triassic rocks of Germany, and were fairly common for the rest of the Triassic.

A single skeleton of the oldest lizard, called *Megachirella*, comes from rocks, also of Middle Triassic age, in the Italian Alps, but apart from this, broken bits of jaws and other skeleton fragments are all that have been found so far of the early lizards. The first fossils of snakes are not found until quite a lot later, in Early Cretaceous rocks. Amongst these, *Najash* still had small back legs. *Tetrapodophis* is the appropriate name given to a controversial Brazilian fossil. It had a very long thin body, a large number of vertebrae, and broad belly scales, all like a snake. But it also had four legs, each very small but with a complete set of bones, digits, and claws. Some palaeontologists think *Tetrapodophis* is truly an ancestral-grade snake, but others are doubtful because the skull is not particularly like that of snakes. What its legs were used for is not certain. Perhaps they helped in digging a burrow, or for grasping captured prey. Its actual diet seems to have been typical of a snake because the presence of some small vertebrae preserved inside the body cavity tells us that they ate other vertebrates.

During the rest of the Mesozoic Era, lepidosaurs underwent a modest evolutionary radiation. Rhynchocephalians declined after their promising start. The lizards and snakes on other hand did much better, and many of the modern groups evolved, although their remains are still not found all that commonly. In today's communities, these smaller reptiles play a distinct role as small-bodied, low-energy-consuming animals that are mostly active during the daytime. No doubt they occupied a similar place in the Mesozoic community. Whatever the reason for their rather limited diversity during this time, those that survived the mass extinction and their descendants took advantage of the new conditions to evolve into the rich variety with us today.

The origin of the chelonians, is more of a mystery. As we saw in Chapter 1, even with the new molecular evidence available, it is still not certain which of the other living amniotes is most closely related to them. Nor for many decades did the fossil record help much in solving the riddle: at least five different fossil amniote groups were put forward by different authors as ancestral to turtles. Of these, one contender all along is called *Eunotosaurus* (Figure 10a), an animal found in Permian rocks of South Africa. It had broad, leaf-like ribs looking as one might imagine the forerunner of the turtle carapace would be, but as no one had found a complete skull, it was not known whether it resembled a turtle at all. However, more complete specimens of *Eunotosaurus* have recently turned up. These show that the skull fitted neatly intermediate between a typical diapsid skull with two temporal fenestrae, and a turtle skull in which fenestrae are absent. The basic problem seems to be solved: chelonians had evolved

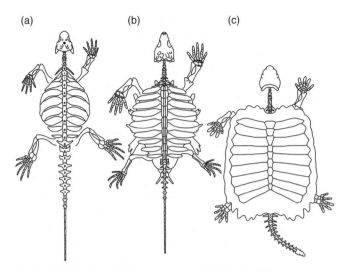

10. **Fossil chelonians:** (**a**) *Eunotosaurus*; (**b**) *Odontochelys*;
(**c**) *Proganochelys*.

from early diapsid reptiles by the Permian, and therefore well
before any of the other modern reptile groups.

Other fossils show subsequent steps in the evolution of the
modern chelonians. *Odontochelys* (Figure 10b) is from the Late
Triassic of China. The shell was not completely formed. While
there was a complete covering of bones, called the plastron, on
the underside of the body, the upper surface only had a single
row of bony plates attached to the vertebrae and ribs, instead
of a complete covering, or carapace. The skull was short like
a chelonian, but it did still have teeth along the jaws. By the end
of the Triassic, turtles like *Proganochelys* (Figure 10c) had evolved
with a fully developed carapace as well as plastron, and the
shoulder girdle and pelvis inside the rib cage in true chelonian
fashion. By this stage, the essentially modern turtle modes of
moving, breathing, and feeding were in place.

Chapter 3
Lizards

More than half of all reptiles alive today are lizards, and with over 6,000 species they count as one of the most successful groups of all land vertebrates. Ranging from within the Arctic Circle in the north to the southernmost parts of South America, on seashores, in deserts and forests, and up 5,000 m high mountains, lizards are truly worldwide in distribution, although like all reptiles they are most diverse and abundant in the tropics. The classification of lizards proved to be a knotty problem until evidence from the DNA sequences of genes started to be used. This provides so much evidence that with its help we now have a pretty clear idea of how the modern lizards evolved. Surprisingly, the first ones to have diverged from the main stem were the dibamids, a rather unfamiliar family of small, tropical lizards that have lost their legs and adopted the specialized burrowing habit. The next were the geckos that include the wall-climbing, fly-catching little animals known to all who have spent time in Mediterranean or tropical climates. Evidently right from the beginning of their modern radiation, more specialized kinds were evolving from what we think of as 'typical' lizards.

Typical lizards

The majority of lizards still have the general appearance of their amniote ancestors. The body is long and slender, and ends in a very long tail. Running is achieved by the combination of throwing the body into lateral waves while the feet provide purchase, and striding movements of the four rather spindly legs that stick out sideways from the body. The feet have long digits and are placed out to the side rather than below the body, making the animal very stable. At rest, the weight is supported by the belly on the ground rather than by the limbs. Despite retaining this basically ancestral mode of running, lizards are extremely fast and agile. Their top speed is about the same as the top speed of a similar sized mammal, and their ability to chase and capture fast-running prey is impressive. In fact their hunting prowess is limited more by a lack of stamina for sustaining a chase than by the nature of their limbs.

Most lizards have a unique specialization of the vertebral column called *autotomy*. One or more vertebrae near to the base of the tail have a special plane of weakness so that when the muscles around it contract it fractures, and the muscles and blood vessels around the break instantly contract to stop blood loss. It is an ingenious means of escaping from a predator such as a bird of prey that has grabbed the individual by the tail. In fact the breakage is so easy and harmless that the tail is often shed as soon as the owner feels seriously threatened. Once free, the tail continues to wriggle, attracting the predator's attention and giving the lizard time to get away. A number of species, especially amongst the skinks like the aptly named cyan-tailed and red-tailed skinks, have a very brightly coloured tail which increases the effectiveness of autotomy by making the shed tail even more likely to be the part that catches the predator's eye. After it has been lost in this way, the tail does regrow, although it never reaches the same length that it had been. Nor, for some reason, does it redevelop the specialized autotomy vertebra, and so this is a once in a lifetime way of escape.

Far and away the majority of lizards are predators, especially of insects, spiders, and other invertebrates, although larger species readily catch and consume vertebrates such as frogs, and small birds and mammals. The way in which lizards feed involves a specialized mechanism called *cranial kinesis*. We are used to the idea that, apart from the jaw opening and closing, a skull is a rigid box-like structure, but in lizards there are other parts of the skull that can move independently. The details vary in different lizards, but the main movement is hingeing the upper jaw and its teeth up and down at the same time as the lower jaw opens and closes. This increases how widely the mouth can open, and how quickly and strongly the jaws can snap shut on the prey (Figure 11). Another common action consists of moving the upper jaw and its teeth backwards and forwards, which helps to manipulate the captured food in the mouth ready for swallowing. Less specialized feeders such as the iguanas have limited cranial kinesis, and their tongue is an important part of food capture. It is a muscular

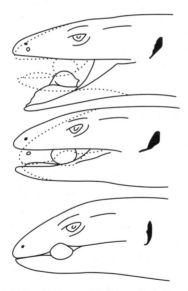

11. **Lizard cranial kinesis. From top to bottom, the mouth-closing sequence showing the movement of the upper jaw independently of the lower jaw.**

structure in the floor of the mouth that can be extended forwards outside the mouth to catch a small item, such as a worm or beetle. The prey gets stuck to it and dragged back into the mouth. In contrast, lizards with more developed cranial kinesis, such as the monitors and whiptails, capture their food using only the jaws and teeth. On the other hand, some specialist feeders, such as the burrowing amphisbaenids and the arboreal chameleons, lose the cranial kinesis altogether.

Those lizards relying exclusively on cranial kinesis for food capture no longer need the tongue for manipulating the food in the mouth. Instead, their tongue is long and slender with a forked tip, and is an important part of the lizard's sense of smell. It is protruded from the mouth and samples the air and the ground for scent molecules. These are passed by the tips of the tongue into a pair of openings in the roof of the mouth that lead to the special olfactory organ called the *vomeronasal* organ. Olfaction is an important means of detecting prey, and in many species plays a role in recognizing their own and other individuals' territorial boundaries, and for detecting potential rivals and mates.

Vision is acute in the typically diurnal lizards, where it is essential for catching live prey such as fast-moving insects, and even grabbing flying insects out of the air as they pass. Their colour vision is also excellent, better in some ways even than that of humans, because as well as discriminating between the three primary colours that we do, some lizards' eyes also have receptors sensitive to ultraviolet light. It is therefore no surprise that colour plays a more important role in the behaviour of lizards than in any other group of reptiles. Some species display extraordinarily conspicuous vivid colours and patterns to attract mates, even at the risk of increasing the chances of their being caught by a predator. For example, the garishly multi-coloured male of the Augrabies flat lizard of South Africa combines a bright blue head, greenish-blue front trunk, yellow front legs, orange hind legs and trunk, black belly, and tan and orange tail, not to mention a UV-coloured throat invisible to us. The female, in contrast, is mostly dark brown with cream stripes. Other species use colour as a form of aggressive display towards rival males. Larger, older males of the European sand lizard *Psammodromus*, for example, develop orange heads to signal their dominance over the smaller, younger males, which helps to avoid physical conflict.

Camouflage using colour, and sometimes body shape as well, is common and can be extraordinarily effective. The lichen-bark gecko of Madagascar has light grey blotches on a black background that make it almost invisible on a lichen covered tree trunk. Even more remarkable is the leaf-tailed gecko (Figure 12a), whose light brown colour coupled with a twisted shape and what look exactly like leaf veins is virtually indistinguishable from a dead leaf. Another lizard that matches its background closely is the flattened, sandy-coloured thorny devil or moloch (Figure 12b) that subsists solely on ants in the Australian desert. This species is unusual in other ways. It is covered in sharp, robust spines and also has a sort of false head on the back of its neck that it presents to a potential predator. This tough structure safeguards its real head from attack.

With so many different species, living in such a wide range of climates and habitats, it is not surprising that there is a considerable variety of patterns of reproductive arrangements amongst lizards. To generalize, most species are described as *polygynous*, meaning that one male mates with several females. The male is usually territorial, holding and defending a particular area by a variety of visual displays, such as the dramatic change from light grey to bright blue of a rainbow lizard when it meets a rival, or the wrestling matches, heads and front legs locked, of monitors. The larger the territory a male successfully holds, the more females he is likely to have. Fertilization, as in all reptiles, is by insertion of the male penis into the female cloaca, but unique to lizards the penis is paired. Only one of the *hemipenes* is used at a time and no one understands why there are two. The number of

12. Camouflaged lizards: (a) the leaf-tailed gecko; (b) the thorny devil or moloch.

eggs the female produces varies from a single large one as in the blind lizards, to more than fifty in some monitors. They are laid in a simple nest or cavity that she carefully constructs beforehand in a safe place, such as rotting vegetation or a crevice between rocks. Here the eggs are left to develop and hatch without any further care from the mother. A small number of lizards do show a degree of maternal care. The prairie skink, for example, coils itself around the eggs to protect them from predators and keep them moist, and some monitors return to the nest at hatching time to assist the emergence of the young.

About 20 per cent of lizards are viviparous; the mother keeps the eggs in her oviduct where they develop until the young are born as miniature adults. Viviparity has evolved many times, sometimes separately in different closely related species, and it occurs in at least some members of most lizard families. One of the advantages of bearing the young live is that the developing embryos can be kept at a higher temperature by the mother basking in the sun. This explains why there is a greater proportion of viviparous lizards living in temperate rather than tropical regions. A typical example is the common European lizard, appropriately called *Lacerta vivipara*. Other benefits are protection of the developing young from desiccation and predation. The amount of additional nutrition that the mother provides to the embryo within her oviduct varies. At one extreme, she does little more than keep the normal eggs in a safe place until hatching, and the embryos rely entirely on the yolk inside their shell. Other species develop a connection between the oviduct wall and the developing embryo, creating a sort of *placenta*. It may be quite simple, or much more extensive such as found in the American mabuyas skinks. In these, the eggs are tiny and all their requirements are provided through the placental tissues from the blood vessels of the wall of the mother's oviduct.

Strangely, about thirty lizard species are *parthenogenetic*, which means they consist only of females which breed without any males.

Parthenogenesis probably arises now and again because the size of the population increases more quickly when all the individuals, not just half of them, produce eggs. This could help colonization of new areas. But the genetic similarity of all the mother's offspring means there is scarcely any variation between them, and so the population as a whole becomes less able to cope with changes in the environment. For this reason, parthenogenetic species usually go extinct before very long.

Even amongst what we have called 'typical' lizards, there is a great deal of variation in habitat, food preferences, and so on. Many lizards, however, have evolved even more extreme modifications to the standard body form and mode of life, and have taken up a variety of specialized habitats. Exquisite adaptations for burrowing, for moving within sand, for swimming, and for tree dwelling have evolved in members of numerous groups. Some lizards became large and hunt down and consume other vertebrates; others developed a fermentation chamber in their gut that is full of bacteria to break down plant food. We'll take a look at some of these more specialized species in the rest of the chapter.

Limbless lizards

Legs are fine for running over hard, reasonably clear ground and for climbing trees, but can be a hindrance for an animal that spends most of its time burrowing, or living deep in the forest litter. The size of the legs of many lizards has reduced, and on more than twenty separate occasions legs have effectively become lost completely. The dibamids, for example, are small, rather worm-like lizards completely lacking any trace of front legs. The back legs are also missing in the females, but a small, flap-like trace of them still exists in the males, where they are used to assist copulation. The common name for dibamids is blind skinks, because their eyes are also almost non-existent, and nor do they have an outside opening for the ear. They are therefore well adapted for a burrowing life and are found in the soils, dead trees,

Reptiles

and deep leaf litter of the forests of Mexico and South East Asia, where they seek out a diet of insects.

Dibamids were the earliest branch of living lizards to have evolved from the main stem, and so may have been the first to reduce their legs and adopt this new mode of life. However, they were followed by members of several other groups. The true skinks are the largest and biologically most diverse family of lizards, with over 1,400 species living throughout the world, from temperate to equatorial regions. They have also managed to occupy many islands, and are particularly abundant in Australia. Many skinks burrow in loose soil and sand, and amongst these the whole range of reduction and total loss of limbs is found in different species. Why this particular group of lizards is so prone to limb reduction is not clear: perhaps it is simply that they already had a generally cylindrical shape that predisposed them to burrowing. For example, the sandfish (Figure 13a) lives in North African and Arabian sandy deserts, where it is adept at what is called *sand-swimming*. The head is shaped like a wedge, and the limbs are only about half the size of a normal lizard's. It can effectively swim just below the surface of loose sand, using lateral undulations of the body as if it were in water, and it can move as fast as two body lengths per second, despite the considerable resistance of the sand.

The flap-footed lizards of Australia belong in the family Pygopodidae. They are so named because they lack front legs altogether, while the back legs consist of no more than a tiny pair of scaly flaps. Like sand swimmers, flap-footed lizards move by lateral undulation, but in their case it is through thick grass, often well above ground level. They are even capable of using the muscular tail to jump up into the vegetation to start with.

The most highly specialized burrowing lizards of all are the amphisbaenians (Figure 13b), all but one of which are without any external trace of limbs. (The exception is appropriately enough

13. Limb reduction in lizards: (a) a sandfish; (b) an amphisbaenid.

called *Bipes* (two-footed), which lacks hind legs, but has quite large front ones.) The common name for these tropical and Mediterranean animals is worm-lizards, and they do look extraordinarily like earthworms. The long, cylindrical body is marked by closely spaced annular grooves around the body, like the segments of an earthworm, and it ends with a blunt head and tail. Many species even have a worm-like pinkish colour. A worm-lizard constructs its own burrow by using the compact, strongly built head to dig and push through the soil, forcing it aside. The skin is very loosely attached so that with the help of the friction provided by the annular grooves, the body can move forwards by a sort of concertina-like motion without having to use the lateral undulations typical of most lizard locomotion. The way amphisbaenians feed is equally well adapted for a burrowing existence. Their eyes are small and skin-covered, and they do not have an eardrum. However, their ears are very sensitive to low-frequency sound carried through the ground, which is the main way they detect their prey. This consists of insects such as termites and beetles that they come across, capture, and drag down their burrow to eat at leisure. Having evolved such a strongly built head for digging, cranial kinetism is no longer possible, and in its place amphisbaenians have strong teeth and a very forceful bite.

Aquatic lizards

There are certain snakes that never leave the sea, turtles that only emerge from the sea to lay eggs, and the saltwater crocodile which spends virtually all its life there. It is strange, therefore, that no lizards have ever become adapted for long-term marine life since the mosasaurs of the Cretaceous that we met in Chapter 2. Lizards are certainly capable of swimming. Several of the varanids, notably the Nile monitor, do spend a good deal of the time in fresh water, but have to emerge regularly to bask in the sun. The water dragons of Australia live in and around riverside vegetation, and if they are threatened, they immediately leap into the water, swim to safety using a slightly flattened tail, and can stay submerged for up to an hour. An extraordinary kind of association with water is found in the basilisk, named appropriately the Jesus Christ lizard in Central America for its ability to run on its hind legs over the surface of the water. The back edges of the toes have fringed scales that retain a pocket of air beneath the foot, so that it can easily be lifted off the surface for the next step. This is a highly effective means of escaping from predators, and like the water

dragon a basilisk can remain under water for a while, until
the danger has gone away.

The closest approach to fully marine life is made by the renowned
marine iguanas of the Galapagos Islands. Up to half a metre long
plus the tail, these spectacularly black lizards, ornamented by a
frilled crest along the head and back, were described by Charles
Darwin as 'imps of darkness'. Like many iguanas, they are a
herbivorous species, but uniquely they feed only on seaweed.
They live in large colonies on the rocky shores of the islands,
basking in the sun. From time to time, they dive down a metre
or so deep into the sea to forage in the intertidal zone, using the
flattened tail and webbed feet to swim, the strong claws to cling
onto the submerged rocks, and the blunt teeth to collect food.
This period is difficult for them because of the low temperature
of the water, but they avoid too great a drop in the body
temperature by reducing the heart rate and the flow of blood to
the skin. On scrambling back onto the shore, they rapidly warm
up once more by basking, aided by the almost black colour of
the body, which absorbs heat faster than would a lighter colour.

Arboreal lizards: chameleons

We can watch many species of lizards climbing trees, running
up the trunk to hide amongst the branches to escape predators,
or feeding on arboreal insects. Their long limbs, clawed toes,
and low-slung body make climbing vertically upwards and
downwards on rough bark a simple matter. Most of them do not
have any particular adaptations for this, but one lizard group,
the chameleons, are so highly specialized for life in the trees that
they are the most bizarre lizards of all (Figure 14). To start with,
chameleons are extremely adept at standing and walking along
thin branches. Each foot consists of two pads, one made up of
three toes and the other of two. These oppose one another either
side of the branch, giving the foot a very strong grip. Then the

14. A chameleon in the act of catching an insect.

legs, instead of extending out sideways as in other lizards, are held almost vertically so that the left and right feet are close together under the body, grasping the same slender branch. Stability is enhanced even more by having what is called a *prehensile* tail, which is one that can curl round and grasp a branch as if it was a fifth limb. The actual movement of a chameleon in a tree is very slow and deliberate, and the animal spends most of its time virtually motionless as it awaits the approach of its prey. Even the head can be kept absolutely still, because the eyes are carried on little turrets and can swivel independently of each other. So remarkable are they, that each eye can rotate through 180° horizontally giving complete all-round vision, and 90° vertically leaving only the regions above and below the body out of sight. The brain coordinates the messages it receives from the two eyes and converts them into a 3D image that lets the chameleon judge accurately how far away an object is. The eyes themselves are extremely sensitive, with many more light-detecting cells than the human eye, and furthermore they can work like a telephoto lens of a camera, to magnify the image on the retina.

Armed with this superb equipment for detecting the prey without disturbing it, and estimating exactly how far away it is, the next thing is to capture it. The means for doing this are no less incredible. The tongue, which normally lies contracted in the floor of the mouth, can be projected an enormous distance by special muscles surrounding and squeezing it. It can extend as much as the body length of the chameleon in a matter of milliseconds. The tip of the tongue is sticky, and has little muscular flaps that also help to grasp the prey, before the tongue is pulled back into the mouth. For most chameleon species, the prey is insects and spiders, but larger members of the family, which can be up to 30 cm in length not counting the tail, happily capture small birds and mammals that come within range of the tongue.

Coloration plays a particularly prominent part in the life of chameleons, and their ability to change colour to match their background is legendary. Sometimes this is in response to a predator, and when threatened the skin may rapidly turn darker. However, it is a bit of a myth that they can change their pattern of coloration to match the background. Chameleons generally live a solitary life, and most colour changes are for signalling between individuals, indicating such things as aggressiveness in males, courtship rituals, and sexual readiness in females. A male panther chameleon, for example, turns from a drab brown or green to a bright red and yellow pattern when meeting a rival male in his territory, indicating a readiness to fight. Fighting is common and consists of much pushing, snapping, grappling, and hissing until one of them gives in and slowly walks away. The East African Jackson's chameleon even has a trio of forward-pointing horn-like extensions on its head. Rival males struggle with one another by interlocking the horns like a pair of fighting stags. During courtship, a male chameleon turns on his brightest colours, even including eye colour, in his attempt to attract the female. The female may respond by her own colour pattern, though generally a good deal less bright than her partner's. Between them, chameleons use in their behaviour virtually the whole of the visible spectrum from red to purple, as well as darkening and lightening the hues, although each species has its own particular distinctive pattern.

Wall lizards: geckos

With almost 1,000 species and a worldwide distribution, geckos are a very successful family of lizards that live in a variety of habitats, including human habitations. Their most distinctive feature is the structure of the foot, which give them an amazing ability to adhere to and move over smooth, vertical surfaces, and even the underside of horizontal ones, in their quest for insects. Each toe is expanded into a pad (Figure 15a and b), and if you look at one of these pads from the underside, you see a row of fine flaps called lamellae crossing from one side to the other

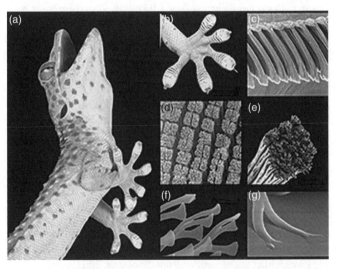

15. The foot structure of a gecko: (a) the whole limbs; (b) the toes with laminated pads; (c) enlarged view of lamellae; (d) surface view of setae; (e) a single seta showing spatular bundle; (f) individual spatulae; (g) contact between a single spatula and the surface.

(Figure 15c). At a higher magnification, each lamella is seen to be made up of millions of minute hair-like structures called setae (Figure 15d). At the very high magnification of an electron microscope (Figure 15e), each seta shows up as a bundle of hundreds of little projections, called spatula, each one a flat triangle which is less than a thousandth of a millimetre across (Figure 15f). There is a total of around one billion spatula over the four feet, and the result is that the total surface area in close physical contact (Figure 15g) with the surface the chameleon is walking over is huge. The total attraction force that exists between individual molecules of the feet and the surface is sufficient to support the weight of the body, even on the smoothest of vertical and overhanging surfaces. This extraordinary clinging ability was originally an adaptation for life amongst rocks, but the way that some gecko species have adapted this habit for climbing vertical walls and crossing the ceilings of human buildings in their quest to catch flies is familiar to anyone who has lived in the tropics.

Unusually amongst reptiles, most geckos hunt at night, and can remain active even with a body temperature as low as 12°C. The eyes too are adapted for night-time. The pupil can be reduced to a vertical slit for daytime use, but expanded into a complete circle to allow in more light during darkness. Another feature of many geckos that is associated with their life at night is that they are the most vocal of lizards, indeed of all reptiles. The barking geckos, for instance, produce a series of clicks and growls in chorus as they busily defend their territory and seek out their mates.

Airborne lizards

True powered flying has only evolved three times in vertebrates, the extinct pterosaurs of the Mesozoic that we met in Chapter 2, the birds which are descended from dinosaurs, and the bats amongst the mammals. But the ability to glide or parachute from tree to tree, or down to the ground without harm, is much more common. All it requires is some means of increasing the area of the body, and the ability is found in a few frogs, snakes, and mammals. Amongst the lizards, the flying gecko, *Ptychozoon*, has a flap of skin down each side of its body that is spread out when the four legs are stretched apart. The head, tail, and feet are broad and flat. The animal usually rests on a tree trunk, where it is very difficult to see, and from where it can leap into the air to escape a predator, glide towards another tree, and catch hold with its toe pads.

The flying dragon, *Draco*, is an agamid lizard that is even more specialized for gliding. It too has a flap of skin down each side, but here it is supported by extremely long ribs that can be extended outwards like a wing, or brought in to the sides of the body. Another flap of skin behind the head helps to control the direction of flight. From a perch high up a tree, a flying dragon can glide in a well-controlled way for as much as 60 m, with relatively little loss in height. As it approaches a new tree trunk, it swoops up and stalls at just the right point to make a gentle landing. Then it rapidly clambers up the new tree to gain enough height for the next glide.

Large predators: monitor lizards and the Gila monster

The monitor lizards make up the family Varanidae, and include in their number the largest lizards alive today, and also the largest that have ever lived. The giant sea lizards of the Mesozoic called mosasaurs (Figure 9d) were members, as was *Megalania*, the largest land-living lizard of all time that was still living in Australia until about 25,000 years ago. It was more than 6 m long, weighed over half a tonne, and would easily have caught and eaten quite sizeable mammals such as small kangaroos and wombats. While a lot smaller than *Megalania*, the largest modern varanid is the Komodo dragon *Varanus komodoensis*, which is still a highly impressive 3.3 m long and 150 kg in weight. It is only found on a few Indonesian Islands such as Komodo, after which it is named, and probably owes its continued survival to the introduction into the islands of domestic animals such as deer, pigs, and goats which constitute its diet. Unusually amongst reptiles, Komodo dragons hunt communally. Their sense of smell, as in all varanids, is particularly acute, and the tongue is long and forked, and is used to sample the ground and the air for the scent of prey. With relatively long legs, they can run as fast as a mammal, although their typical hunting strategy is to lie alongside a game track and ambush the prey. They use their strong, sharp claws to bring it down, and the bite from the sharp, serrated teeth introduces a toxic saliva that adds to the effect of blood loss and shock to cause fairly rapid death of the victim. Cranial kinesis of the skull is especially well developed in varanids, which helps the Komodo dragon to swallow large chunks of flesh torn off the prey. As well as live mammals, carrion is readily taken, and very occasionally humans are killed and eaten. However, the combination of the low metabolic rate of a reptile and the large size of most of their prey means that Komodo dragons actually only need to feed once every two or three weeks.

If not as large as the Komodo dragon, many of the other sixty or so species of varanids are still relative giants amongst lizards. The Nile monitor of Africa (Figure 16), for example, can reach over 2 m in length. It is an excellent swimmer as well as runner, spending much of its time in water, and including fish as well as mammals, birds, and frogs in its diet. The crocodile monitor of the New Guinea coastal forest is another large varanid, also growing to over 2 m including a very long tail. It is a habitual tree dweller, where it feeds mainly on birds.

The Gila monster and closely related Mexican beaded lizard make up another group of specialist predatory lizards, although in their case they are only up to about half a metre in length. They both occur in desert regions of southern North America, where they spend much of the time beneath rocks or in burrows to avoid

16. Nile monitor, showing the extended tongue.

overheating. Their most unusual feature is that they are the only truly venomous lizards. The venom is produced in modified salivary glands as in snakes, and is delivered via the grooved, curving teeth when the animal bites. It seems to serve a defensive rather than a food capture role, because their main food is actually the eggs of both ground-nesting and tree-nesting species of birds, and of other reptiles. With the help of strong, sharp claws, they also find and consume young mammals in their nests.

Herbivorous lizards: iguanas and others

Very few lizards have adopted a herbivorous diet, only around 1 per cent of all species, although many are more or less omnivorous, eating a significant amount of plant food in addition to their animal diet. The difficulty for a lizard is that they are unable to digest the cellulose that makes up plant cell walls, and therefore have to evolve a special part of the gut to contain microorganisms that can. Furthermore, plant material, especially the leafy parts, is quite low in nutrition so large amounts have to be consumed and most reptiles do not have suitable teeth for this. Also, being ectotherms that cool down at night, for long periods the body temperature is too low for very rapid digestion.

The iguanas are the only lizard group whose members are practically all herbivores, and one specialist case, the seaweed-eating marine iguana of the Galapagos Islands, has already been described. Most other iguana species also live in the Americas, although a few have dispersed into South Pacific islands such as Fiji and Tonga, probably carried on rafts of vegetation from South America some time in the remote past. They have evolved several adaptations for successfully surviving on plant food. The body of iguanas is usually on the large side, for example a male of the widespread green iguana reaches about 40 cm in length not counting the long tail and weighs up to 4 kg. Larger animals have a relatively lower metabolism, which for a herbivore is an advantage as it means that they need relatively less food. The actual food consists mostly of leaves, which are cropped by a set of sharp, blade-like teeth set along the jaws and operated by a strong bite force. Once swallowed, the food eventually reaches the colon, which is part of the large intestine. It is large and ridged to increase its surface area, and contains the bacteria and other microorganisms that break up the cellulose into sugars that the animal can absorb.

There are very few other more or less completely herbivorous lizards. *Uromastix*, the spiny-tailed lizards of Africa and Asia, feed mainly on green plants but occasionally catch and eat invertebrates. Their growing young in particular need plenty of protein, and consume a lot of insects. However, many lizard species do take a certain amount of plant food in their diet, mainly in the form of flowers and fruit which are easier than leaves to consume and digest. A few lizards specialize largely in feeding on nectar, and one particularly remarkable example are the day geckos of Mauritius. Not only do they benefit from this highly nutritional food source, but certain plants also benefit because the geckos are their main pollinators, so much so that the plants have evolved brightly coloured blue or red pollen to attract the lizards to them.

Tuatara: the 'lizard' that is not a lizard

On a few rocky cliff islands off the coast of New Zealand, and in a small protected mainland sanctuary, lives a half-metre-long reptile called *Sphenodon punctatus*, or to give its local Maori name, the tuatara, which means 'peaks on the back'. To all appearances, it is a kind of lizard, with a scale-covered body, and sprawling limbs and tail of typical lizard proportions. The serrated crest down the middle of the back is even similar to those found in many iguanas. But *Sphenodon* (Figure 17) has several features that tell us that it is no more closely related to lizards than it is to

17. *Sphenodon*, the tuatara of New Zealand.

snakes. It is a member of the Rhynchocephalia, the sister group of the squamates, and a much more diverse and abundant group 200 million years ago than were the Squamata. However, apart from this one branch that survived in the isolation of New Zealand, rhynchocephalians had completely disappeared by 70 million years ago.

Amongst its characters are the absence of such lizard features as a double penis, and cranial kinesis of the skull. The teeth are also different from those of lizards. In *Sphenodon* there is a double row of upper teeth that are fused together to create two parallel, continuous serrated blades that bite either side of a similar single blade made up of the fused lower teeth. Using this very effective cutting organ, tuataras live on a diet of invertebrates, and also the eggs and chicks of nesting seabirds. Their New Zealand environment is a cool one, and even more so because tuataras are mainly active at night, and the body temperature is correspondingly very low, usually around 10–15°C. They shelter in burrows in soft sand during most of the day, which they either dig for themselves, or frequently take over from nesting burrows of petrels and shearwaters.

The low metabolic rate also explains why their reproduction and development are extraordinarily slow. The female only lays eggs once every three or four years, the eggs take over a year to develop and hatch, and the juvenile needs around ten years to become mature. On the other hand, life expectancy of an individual *Sphenodon* is huge, for it is likely to live for as much as a century.

Chapter 4
Snakes

No group of animals has attracted more aversion and fear in people than snakes. It comes from a combination of their potentially fatal venomous bite, their creepy furtiveness as they lie in wait and appear by surprise, and the mistaken belief that they are unpleasantly cold and slimy. But from our point of view, snakes are to be admired as possibly the most superbly adapted hunting organisms in the natural world. A typical snake does not need to chase and tussle with its prey like other predators do, even though the prey itself can be so large that the snake only needs to feed once every few weeks.

The roughly 3,500 species of snakes are really a group of limbless lizards, although they are so distinctive, with so many unique features, that they are placed in their own reptile subclass, Ophidia or Serpentes. There is uncertainty about exactly where on the lizard evolutionary tree they fit. For a long time they were believed to be related to the somewhat similar limbless, burrowing worm-lizards (amphisbaenids) and blind skinks (dibamids), described in Chapter 3. But molecular evidence has shown that the similarities are actually due to independent, parallel evolution rather than to sharing a common limbless ancestor. Instead, the snakes evolved from deep within the lizard evolutionary tree, with no particularly close living relatives. The least modified of the living snakes are the blind snakes, a group given the name Scolecophidia, which means worm snakes. They are small, burrowing snakes whose eyes are greatly reduced, as their common name implies. They do not have the broad belly scales of all the more progressive snakes, and feed on small items of prey such as ants and termites. The rest of the snakes are members of the far more diverse group called Alethinophidia, which means 'true' snakes. These do have wide belly scales, and their skulls are specialized for feeding on large prey.

Today the snakes, like the lizards, have a worldwide distribution and occupy a great many habitats, from deserts to rain forests and seas, even including a small number of species that survive above the Arctic Circle. But it is the tropics that hold the greatest richness of diversity. Unlike lizards, snakes show very little variation in their general body form. All the living species completely lack limbs, apart from a pair of minute horny spurs alongside the cloaca in boas and pythons, a vestige of the hind limb that is used to stimulate the female in copulation. The number of vertebrae in the vertebral column is huge, 200–400. Each one attaches to the next by a strong ball-and-socket joint which allows bending, and four interlocking bony contacts to make sure they are not easily disarticulated. Snake scales are overlapping and flexible, and the single row of very wide scales along the belly found in most snakes are in contact with the ground, helping to increase the traction.

There is a certain amount of argument about how the snake ancestor lived. Most zoologists believe that they first evolved as underground burrowers, which explains why their limbs were lost, and also why their eyes are different from those of other reptiles. A more unlikely suggestion is that snakes began in the sea, because many modern snakes are such excellent swimmers.

How snakes feed

The main aspect of snake biology that accounts for their great success is the way they acquire and ingest their food. Uniquely within the animal kingdom, they can catch, kill, and swallow whole very large living prey, often larger even than the size of their own head. This in turn gives them the additional advantage of only needing to feed at relatively long intervals, as much as two or more weeks, and if necessary a great deal longer.

Cranial kinesis of the skull has already been described in lizards, where it helps in capturing the prey and moving it around the mouth ready for swallowing. Snakes inherited this ability from their lizard ancestors but took it to extraordinary lengths. The simplest mechanism is found in the blind snakes, which do not take large prey, but have specialized in eating ants, termites, and other insects in very large numbers. Their eyes being useless, they detect their prey by smell and collect it in the mouth by the action of the upper jaw. The upper jaw bone of each side, called the maxilla, is loosely attached to the rest of the skull and carries two large teeth. As it swings rapidly forwards and backwards, the insects are raked into the mouth from where they are swallowed directly.

We are not sure whether this simpler feeding mechanism and smaller prey was inherited from the ancestral snake, or whether the blind snakes later simplified their jaws. But whichever is the case, all the other living snakes have a more elaborate arrangement of the skull bones that greatly increases the size of prey that can be taken in, manipulated in the mouth, and swallowed whole. The two upper jaws are hinged to the skull at the front ends, and the back ends can swing downwards, while the lower jaws can open extremely widely, by as much as 180° (Figure 20c). Added to this, the front tips of the lower jaws are not firmly attached to one another as in other reptiles, but are connected by elastic tissue, so that the two jaws can spread wide apart. By means of these movements of the bones, a huge gape can develop that is directed forwards, and lets the snake ingest an item of food larger than the size of its whole head (Figure 18). Another part of the cranial kinetism is that the left and the right jaws can move independently of each other. Once inside the mouth, the food can be worked

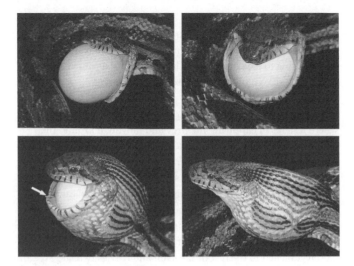

18. Sequence showing a snake swallowing a whole egg. The white arrow shows the tip of the right mandible supporting the elastic ligament connecting the jaws.

backwards and pushed down the throat by the left jaws then the right jaws alternately grasping, pulling, releasing, and moving forwards ready for the next pull. It is similar to how we pull on a rope, hand over hand.

All biological adaptations, especially the more extreme ones, generally have knock-on consequences that have to be compensated for by other evolutionary changes, and the kinetic feeding mechanism of snakes is an excellent example. The main problem it raises is due to the weakening of the skull, as individual jaw bones became loosely attached to one another and to the rigid central braincase. If the snake attempted to ingest violently struggling prey, the skull would be quite unable to contain it without damage. Therefore the victim must first be completely immobilized. One way to do this is called constriction (Figure 19). The snake coils its body around the prey and tightens the coils. Death follows quickly, due

19. An anaconda constricting a caiman.

to restriction of the blood flow or, in the case especially of mammal prey, by asphyxiation as the pressure on its chest prevents breathing. Constriction is the means used by most non-venomous snakes such as boas and pythons, and many of the mildly venomous ones such as the garter snakes also frequently use it.

It is perhaps surprising how few animals use such an efficient hunting method as injecting a fatal dose of a highly toxic substance into their prey. A small number of spiders, centipedes, and insects, the jellyfish, a few fish species, the Gila monster amongst lizards, and possibly one or two shrews is the total. But to these we must add about 600 out of the 3,500 species of snakes. In fact many, perhaps most of the snakes described as non-venomous do have toxins, but they lack specialized teeth for injecting large amounts. There are about 200 venomous species that are a threat to humans because they use grooved or hollow fangs to inject their venom directly into the bloodstream. This rapidly immobilizes, kills, and even starts the digestion of the prey before it is drawn into the mouth and swallowed whole.

With over 1,800 different species, the family Colubridae contains more than half of all snakes. They are very diverse in their diets and feeding habits, but most of them are at best only mildly venomous snakes. They are known as the rear-fanged snakes, because the venom is delivered by teeth at the back of the upper jaw. Generally, colubrids cannot bite deeply and effectively enough to inject sufficient venom to harm a large animal. In just a few species, notably the African boomslang that can inflict a fatal bite on humans, the teeth are enlarged and grooved, which makes the venom transfer more effective. Amongst the huge range of prey in different colubrids, the common grass snake, which is an adept swimmer, specializes in frogs and fish. The very fast-moving racers catch mostly lizards and even other snakes, while the appropriately named snail-eating snakes can pull a snail out of its shell using their long, strong front teeth. Many colubrids prefer small mammals such as rodents, killing them by constriction first, before injecting venom to speed up the digestion. Yet another colubrid is amongst the most spectacular feeders of all snakes. The egg-eating snakes of Africa swallow birds' eggs whole, even ones the size of a chicken's egg, which is several times the volume of the head (Figure 18). The egg remains intact as it passes into the intestine, enormously distending the body. Once inside the intestine, it meets a spike of the vertebrae that breaks the egg open to release its contents.

Two families of snakes, called the front-fanged snakes, have separately evolved a more effective way of delivering venom, and consequently they include the most dangerous snakes. They have a specialized pair of fangs towards the front of the upper jaw. The advantage of this is that when the snake opens it jaws widely and raises its head, the fangs point more or less directly forwards ready for a rapid, direct strike at the prey (Figure 20c). In the elapids, for example the mambas and cobras (Figure 20a), the fangs are elongated and hollow, although the maxillae bones which house them are immovable, so the length of the fang is limited by the need to be able to close the jaws. The sea snakes are also

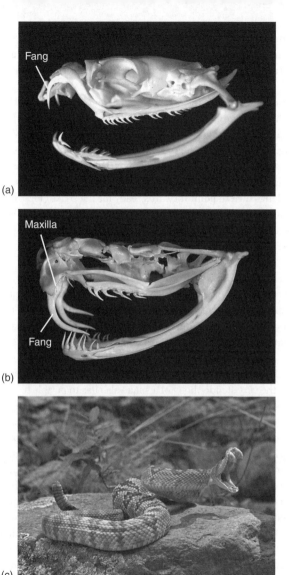

20. Front-fanged snakes: (a) skull of black mamba, an elapid; (b) skull of rattlesnake, a viperid; (c) rattlesnake poised to strike with fangs erected.

elapids. They live in the tropical regions of the Indian and Pacific oceans, where they use their extremely toxic venom to capture other fish, often seeking out their prey in crevices on coral reefs.

The other kind of front-fanged snakes are the viperids, such as rattlesnakes, puff adders, and the South American terciopelo (fer de lance), which have evolved even more effective venom delivery (Figure 20b). When the mouth is closed, the hollow fangs lie horizontally along the roof of the mouth, but when it opens, the maxilla is automatically rotated forwards and downwards so that the fangs turn to point forwards ready for the strike (Figure 20c). This arrangement means that the fangs can be longer, and can therefore penetrate deeper into the body of the prey and inject a greater volume of venom.

The spitting cobras are unusual elapids, because they can eject venom from a hole in their fangs for several metres, as a means of defence. The venom has no immediate effect on the receiver's skin, but the stream is accurately aimed at the eyes, where it causes great pain and can result in permanent blindness.

Swallowing whole prey that is larger than the head requires that the very slender body of a snake also has to be able to accommodate it. This is dramatically illustrated by seeing the bulge in a metre-long snake that has recently swallowed an adult rat, or reading reports of a 4 m long green anaconda that had apparently consumed a jaguar. There are even verbal reports of seeing humans killed by constriction and swallowed by pythons, mostly children but also one or two adults, although this is certainly an extremely rare occurrence. To allow such a large item to enter the gut, the ribs are free below and can rotate upwards and outwards to increase the body's diameter, while the skin is elastic and the intestine highly distensible. Other organs of the body, the liver, paired kidneys, and single lung, are long and thin, and there is no bladder to fill with urine and interrupt the passage of food. The urine is passed directly into the cloaca.

We have already mentioned the advantage to snakes that feed on very large items of prey of not needing to feed very frequently, typically no more than once every week or two, and often they can survive without any food for several months. To save energy in between meals, snakes reduce their metabolic activity, and the temporarily unused digestive organs shrink. But within only six hours of taking in food, a python, for example, increases its metabolic rate some sixfold, and after two or three days organs like the small intestine, liver, pancreas, and kidney have doubled in size, and active digestion is taking place.

Snake venom

The venom is produced and stored in large glands lying alongside the upper jaw. They are modified salivary glands, and it is no coincidence that the venom of snakes also has a digestive role in starting the process of breaking down the prey's tissues, even before it has been swallowed. This is important, because by swallowing the prey whole without chewing it up first, the enzymes of the intestine alone would take a very long time to complete the digestion. This double function of venom, killing and digesting, is reflected in the two main kinds of molecules it consists of. Some are called neurotoxic, and these directly immobilize the prey by preventing the nerves from transmitting their impulses to one another or to the muscles, rapidly leading to paralysis, asphyxiation, heart failure, and inevitable death. Others are digestive enzymes, called haemolytic, that attack the prey's tissues, destroying cell membranes, attacking muscle tissues, and making the blood and cell tissues more fluid to cause haemorrhaging and rapid flow of the toxins through the body.

Any particular species of venomous snake has a mixture of different enzymes. Some are more dominated by haemolytic molecules, others by neurotoxic molecules. There is no simple relationship between types of venom and kinds of snakes, but generally speaking the viperids, such as rattlesnakes, puff adders, and the terciopelo, have mostly haemolytic venom. The elapids, such as the cobras, mambas, and kraits, have a greater neurotoxic content. The question of which are the most dangerous snakes is hard to answer exactly, because it does not depend only on the toxicity of the venom, but also on the effectiveness of the delivery by the fangs, and the general habits of the species. In purely molecular terms, the most toxic venom as measured on mice and on human tissue cultures is possessed by the inland taipan, an Australian elapid. However, it is shy and inclined to flee rather than confront a potential aggressor, and so is not counted amongst the most dangerous species to humans. That accolade is generally accorded jointly to its close relative the coastal taipan, and another elapid, the black mamba of Africa. Both these species turn immediately aggressive when feeling threatened, rearing upwards and striking multiple times in extremely quick succession. In the case of the black mamba particularly, it seems that they never deliver dry bites without injecting venom, as many other snakes do when attacked. Instead they always deliver venom, and even though the lethal dose for humans is only about 15 mg, the snake injects as much as 120 mg, via the longest fangs of any elapid snake. The venom contains a particularly fast-acting enzyme that reduces the viscosity of the tissue fluids to increase how rapidly it spreads through the body. Symptoms of neuromuscular failure start after less than ten minutes, and death follows after as little as thirty minutes. Untreated black mamba bites are virtually 100 per cent fatal, and even with immediate treatment the death rate is about 14 per cent, because of the speed with which the toxic effects appear.

The pathology of serious snake bites in humans varies enormously, not only with the species of snake involved, but also with how much, if any, of the venom is successfully injected. Bites from vipers and cobras tend to lead to immediate severe pain, swelling and bleeding around the joint, and, as the venom spreads, internal bleeding and necrosis of tissues. Death can eventually follow from collapse of the blood pressure and consequent kidney failure.

An elapid bite may not be very painful, but neurological effects like blurred vision and numbness may soon follow, and ultimately death due, for example, to respiratory failure.

The danger to human life caused by snake bites is often exaggerated, though by no means non-existent. In a great many cases, it is extreme fear that causes some of the symptoms of a snake bite, such as faintness. In other cases, such as where no venom was actually injected, bacterial infection introduced by the bite or at an untreated wound is a greater danger. It is impossible to be sure of the figures because many bites are never officially recorded, but it is estimated by the World Health Organization that up to five million snake bites occur each year. Of these, only between 94,000 and 138,000 prove fatal. This small fraction is due partly to the fact that many, perhaps around half of the bites, are dry, meaning that the snake deliberately avoids injecting venom in order to conserve it. In other cases there may not have been sufficient time or deep enough penetration by the fangs to inject much venom. However, the main reason for the low mortality nowadays is the availability and effectiveness of antivenom treatment. All those traditional treatments recounted in adventure tales, such as applying a tight tourniquet, cutting the wound and sucking it, or applying potassium permanganate crystals, have been proved not to work, and even in most cases to cause more harm still. The only immediate first aid now recommended is to immobilize the bitten limb with an elastic bandage, which can slow the spread of the venom via the lymphatic system. Antivenom consists of antibodies against the venom of one or more particular snake species, created in another animal such as a horse by injecting it with a small, non-lethal sample of the venom. After injection, the antivenom starts to act immediately on the patient, although it cannot reverse any damage already caused by the bite, which is why getting to hospital as fast as possible is so important. Also, there is the added complication that some people have a severe immune reaction to the serum itself that may need careful monitoring.

There is a fascinating evolutionary consequence called mimicry resulting from some but not all snakes being venomous. Some venomous species have evolved conspicuous warning coloration, which is easily recognized and subsequently avoided by potential predators that have come into contact with them but survived. The coral snakes of Central America are dangerous elapids and are coloured in bright red, yellow, and black bands. They act as what are called models for certain other unrelated snakes which are not venomous. These have evolved a very similar colour pattern to the model, and by misleading their predators such as snake-eating raptorial birds, they gain the same protection.

How snakes move

A few kinds of living snakes including the boas and pythons still have the tiny remains of the back legs embedded in the muscles, either side of the cloaca. But at most nothing larger than a pair of little claw-like protrusions can be seen externally, and they are used, if at all, only to help position the female during mating. Otherwise there is no trace at all of limbs. Despite this, several different kinds of locomotion have evolved. The very large number of vertebrae, typically 200–300 but as many as 400 in some species, gives the body extraordinary flexibility. The basic way of moving is by lateral undulation of the whole body. Contraction of one side of the body behind the head throws that part into a curve. The contraction then continues like a wave backwards along the length of the body. Meanwhile a second contraction, on the opposite side of the body, starts from the front and passes backwards, then a third, and so on. As long as there is enough friction between the belly scales of the snake and the ground, the waves passing backwards down the body drive the animal forwards and the snake progresses. Watching how helpless a snake is when placed on wet glass shows how necessary is the friction.

Throwing the body from side to side like this might seem a very inefficient way to move forwards, but nevertheless snakes can reach surprisingly high speeds at least over short distances: the black mamba is allegedly the fastest snake in the world, and can reach 20 kph. Most snakes are sit-and-wait hunters and do not normally need to move at high speed. Some snakes, however, do energetically pursue their prey. The Galapagos racer, a colubrid, is famous for the way that groups of them chase hatchling marine iguanas, even jumping over gaps between rocks and up vertical faces in their efforts to get close enough to grab their prey.

This same species further demonstrates its great locomotor versatility by sometimes hunting for fish in rock pools, reminding us that a similar lateral undulation of a long, slender body is how eels swim. Many snakes can swim effectively in just the same way. For example, the European grass snake and the slender North American garter snakes are often seen swimming on the surface of a lake or stream, seeking out frogs and fish near the surface of the water. In South America, the huge green anaconda is up to 5 m long and spends most of its life in the marshes and pools of the tropical rainforests, with only the head showing. Here it waits for passing prey such as a deer, caiman, or duck to pass near enough to be grabbed. The sea snakes are extremely venomous elapids that spend their whole life in the sea, diving below the surface to hunt for fish amongst rocks and crevices. Their swimming ability is increased by side-to-side flattening of the body, and a long, paddle-shaped tail whose beating provides most of the propulsion.

Snakes living in a burrow, whether made by itself, or as often happens taken over from a rodent, cannot move by normal lateral undulation. As long as the burrow, or even a crevice in the rocks, is slightly wider than the snake's body, it can use what is called *concertina* locomotion (Figure 21b). A short part of the body is thrown into tight curves so that the sides press against the burrow wall. This anchors the snake, which then extend its front region forwards at the same time as the region of contraction shifts backwards down the body. A new anchor is then made near the front and the process continues. The burrowing shield-tailed

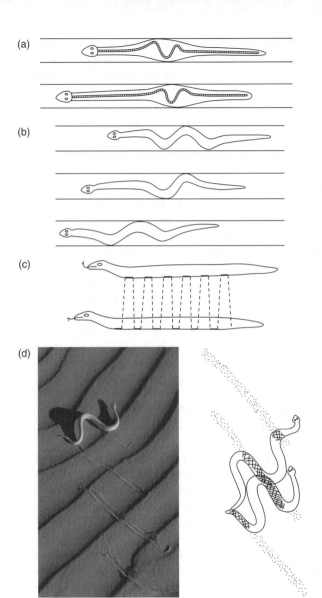

snakes of India use another kind of concertina locomotion (Figure 21a). Here the skin is loosely attached and can easily stretch. The vertebral column within the front part of the body is thrown into a concertina shape, but this makes the body bulge rather than bend, creating a grip on the sides of the burrow. The expanded region of the body moves backwards but maintains the grip, which drives the head forwards. Once the bulge has reached the hind end of the vertebral column, a new bulge forms at the front and the process is repeated. Burrowing in these particular snakes is helped by a strongly built, pointed skull that is driven through the soil as they create their own burrows.

Snakes that live in deserts, for example the sidewinder rattlesnake of North America, often need to move over a surface of loose, hot sand which offers little friction. They employ *sidewinding* (Figure 21d), which is a surprisingly rapid means of progression. As a wave of lateral contraction moves down the snake, most of the body is raised clear of the ground, leaving only two or perhaps three points of contact. The snake progresses in a forwards and sideways direction, relying on the weight of its body acting downwards on the ground to stop backsliding, instead of on friction. Also, much less of the snake's body is in contact with the hot ground, reducing the heating effect.

21. Snake locomotion: (a) concertina locomotion using undulation of the vertebral column to swell part of the body and grip the burrow, as in shield snakes; (b) concertina locomotion using tight undulations of the whole body to grip the sides of a burrow or crevice; (c) rectilinear locomotion, in which short sections of the underside of the body stretch and contract and the belly scales prevent backsliding; (d) sidewinding: the diagram on the right shows two successive tracings of the snake about one second apart, as it moves up the page. The shaded sections of the body are the parts in contact with the ground and the unshaded sections are the parts held off the ground and being moved sideways; the stippled lines are the body prints made in the sand that will be left behind, as can be seen in the photograph of a sidewinding rattlesnake.

Most snakes can move by sidewinding when necessary, and often use it when travelling over wet mud, which does not provide enough friction for normal locomotion.

Another specialist mode of moving is used by large, heavily built snakes such as boas and pythons when stalking prey. It is called *rectilinear* locomotion (Figure 21c) because the snake keeps its body straight, rather than throwing it into sideways waves. Short sections of the belly are contracted and expanded relative to the rest of the body using muscles connecting to the ribs. The broad, overlapping ventral scales act like a ratchet to prevent backsliding. Rectilinear locomotion allows the snake to creep surreptitiously towards its prey, scarcely disturbing the vegetation at all.

Many snakes are excellent tree climbers, and some lead a full-time arboreal life. They move around the branches and lie in wait for their prey, such as small mammals, and nestling birds. The brilliant green emerald tree boa of the Amazon Basin and the very similar looking but unrelated green tree python of Australasia and Indonesia, for example, have a prehensile tail that grasps branches by coiling tightly around them. The underside of the body is flattened and can be shaped to fit the branch along which the snake is draped. More delicately built species, such as the snail-eating snakes and the vine snakes, can extend the front part of the body horizontally forwards towards another branch. This is called *gap-bridging*, and they can hold up to half the length of the body freely extended in this way. Without clawed limbs, climbing up a vertical tree trunk could be a problem for arboreal snakes, and the solution in some species is to have ridged ventral scales that help to hold on to rough bark as the snake ascends.

The mildly venomous colubrid *Chrysopelea* is the flying snake of south-eastern Asia. It is up to about 1.2 metres in length, and has several arboreal adaptations, such as ridged ventral scales, and a flattened belly. However, it has taken arboreal existence even further, for it is capable of gliding through the air from tree to tree for distances as much as 100 m. When ready to launch itself, the snake extends its ribs forwards and outwards. This doubles the width and surface area of the underside, and creates an aerodynamic shape like the wing of an aeroplane. As it throws itself forwards into the air, undulations of the body apparently make it an even more efficient glider, although we are not sure exactly how.

How snakes sense their surroundings

The senses of sight, smell, and hearing of snakes all differ from those of other reptiles in ways that can be explained by the theory that snakes evolved from a burrowing lizard. Well-adapted modern burrowing lizards have very reduced eyes. If snakes evolved from some such an ancestor, then their eyes must have re-evolved from tiny eyes, sensitive to little more than light and darkness. They became the large eyes, able to detect images and colours, which most of them now have. The way in which a snake's eye focuses on objects at different distances is unlike that of any other amniote. The spherical lens is bodily moved backwards and forwards by muscles, instead of its shape being changed by muscles, called *ciliary muscles*, which squeeze around the edges to fatten it. Another striking difference is that snakes' eyes are protected by a tough transparent layer called the spectacle, made up of fused, colourless scales. There are no moveable eyelids, which is why snakes have their sinister, unblinking stare, sometimes said to have a hypnotic effect on their prey.

Snake eyesight is not particularly acute, and is mainly used to detect movement rather than to sense high-resolution images. The ball python is a small African species whose vision has been studied in detail. Like most vertebrates, it has two kinds of light-sensitive receptors in its retina, rods and cones. The rods are sensitive to blue light, which is best for seeing in low light, and they are very densely packed, giving the eye the visual discrimination needed for hunting. Colour vision, as in all vertebrates, depends on the cones. In this snake there are only two kinds. One is sensitive to green light; the other rather surprisingly is stimulated by ultraviolet light, which is light whose wavelength is shorter than our eyes can see. It is not certain why they should have evolved this ability, but many odour molecules reflect UV light, so perhaps the ball python can detect and follow scent paths of potential prey, or the pheromone trail left by a female, using sight as well as smell.

The eyesight of tree-dwelling species, such as the vine snakes, is much better than other snakes. The two eyes work together to accurately judge distances between branches, which is called *stereoscopic*, or binocular vision.

While vision is not especially good in snakes, the same cannot be said for their sense of smell. Their olfactory, or chemoreceptive sensitivity is exquisite, and is another hint of the underground origin of snakes. The long, deeply forked tongue flicks in and out about once a second, picking up odour molecules from the air and the ground. The fork tips pass the molecules through the pair of openings in the roof of the mouth leading into the *vomeronasal* or *Jacobson's organ*. This is lined by receptors, which between them are highly sensitive to a huge variety of odour molecules. Olfaction is the main way in which a snake identifies any particular kind of prey, and is used to follow the scent trail it leaves. It also plays a major role in reproductive behaviour, such as recognizing the pheromones secreted by a female.

Like sight, the way a snake hears differs from that of other reptiles. The latter have an ear drum that picks up sound waves carried in the air and passes the vibrations via a tiny bone called the *stapes* to the sound-sensitive organ, the *cochlea*, inside the skull. Snakes in contrast have lost their eardrums. Instead, sound vibrations are picked up by the rear part of the lower jaw. This is a much less sensitive method of hearing, and is used mostly to detect vibrations carried through the ground. It is true they can use it to detect airborne sound waves as well, but only of a fairly low frequency and high volume. A typical snake can only hear up to about the middle C note of a piano. Nevertheless, sound detection of these lower frequencies is important, because it is the main way a snake detects the nearby footsteps of a possible predator, including an approaching human.

The ability to detect heat is extremely useful to an animal that hunts warm-blooded mammals, and it has evolved in three different kinds of snake. Pythons, boas, and pit vipers such as the rattlesnakes have pits opening onto their face, a single pair in the case of a pit viper and a row along the jaws in the other two. They contain sense organs that are sensitive to the infrared heat rays given off by a warm body, even in total darkness, and they are so sensitive that they can detect a difference in temperature of as little as 0.001°C. In practice this means the snake can detect a mouse from up to a metre away. It can also pinpoint precisely the direction the heat is coming from, by comparing minute differences in temperature detected by the pit organs on either side of the head. These infrared receptors not only provide the snake with the position of potential prey, but also create a sort of heat image of its surroundings, comparable to a visual image. Therefore they play a larger role in the snake's life than just hunting, and can be used for activities like detecting and avoiding predators, and seeking out mates.

Social and reproductive behaviour

Snakes are amongst the most solitary groups of vertebrate animals. In a few species, groups of individuals do come together, but only for such purposes as hibernation in the winter. The most notable example of this is the garter snakes of North America, where several thousand may aggregate in the same shelter for mutual protection against the winter cold. Favoured basking sites may also host a number of individuals of a species, such as the prairie rattlesnakes. But even in these cases, there is no social organization into dominant and submissive individuals, or any particular behavioural signalling amongst them. Nor do snakes defend territory. In fact, for virtually all species, the only social interactions are to do with reproduction.

Mates are recognized mostly by olfaction, because there is very little in the way of visible differences between the sexes of snakes. Courtship typically involves the male chasing the female, along with licking, biting, and trying to mount her. Male pythons and boas may use their tiny hind limb remnants to stimulate her. If the female is receptive, the successful male coils around her to bring their genital organs together to copulate.

As we saw in the lizards, viviparity is also quite common in snakes, especially in species living in the cooler areas of higher latitudes and altitudes, where the mother's basking behaviour raises the temperature of the developing embryo. In North America for example, about a quarter of the snakes living up to 30°N bear their young live, but almost two-thirds of those that live beyond 50°N do so.

The great majority of snakes simply lay their eggs or bear their young in a suitably safe place, and abandon them completely. But there are a few examples of maternal care. The best known are the pythons (Figure 22), where the mother coils herself around her developing clutch of eggs to incubate them. She raises her own body temperature by the remarkable method of shivering her body muscles to generate heat, which keeps the eggs warmer so they develop faster. Of course, this behaviour also protects them from the various birds, mammals, and other reptiles that would happily devour them. Several other snakes, such as a number of pit vipers, guard their eggs during development up to hatching, and the females of the viviparous ones like the cottonmouth stay around for a week or two protecting the newly born offspring. However, the most remarkable case is the

22. Female ringed python brooding her eggs.

horned rattlesnake or sidewinder. The mother bears her live young
in the burrow of a rodent which she has taken over. The dozen or
so young mass together near the opening of the burrow, forming
a sort of plug. But they move around, so that each individual is
sometimes near the entrance in the hot sun, and sometimes
further in where it is cooler. In this way they all keep their body
temperature close to the ideal. Meanwhile the mother stays nearby
to protect them against predators.

Chapter 5
Crocodiles

Compared to the 6,000 species of lizards and 3,500 species of snakes, our third group of living reptiles is very modest indeed. There are only about twenty-five species of living crocodilians (different experts disagree on exactly what the correct number is). Furthermore, these modern ones all have a very similar body form, with the familiar long head with rows of sharp-pointed teeth, heavily scaled, somewhat flattened body, relatively short but stout legs, and massive tail. They all live in a semi-aquatic habitat, mostly around freshwater rivers, lakes, and swamps. A few venture into marine areas too, notably the 'saltie' or saltwater crocodile, which has spread all over the coasts and seas of northern Australia, Papua New Guinea, and as far as the islands of Indonesia and the Indian Ocean. Even the size range of crocodiles is quite narrow by the standards of other reptiles. The smallest is the dwarf crocodile, although this still grows to as much as 1.9 m long. It is a shy species and is only active at night, in the swamps and broad, shallow rivers of the rainforests of western central Africa where it makes its home. At the other extreme, the largest reliably measured crocodile is a 7 m long saltwater crocodile. The Nile crocodile, the most widespread and abundant African species of all, is not far behind with a length often approaching 6 m. As well as body form and size, the geographical range of crocodiles is quite limited too. Almost all of them live within the tropical and subtropical regions.

The American alligator extends into the warm temperate zone of the eastern USA, but only as far northwards as North Carolina. The range of the severely threatened Chinese alligator, which is a close relative of the American one, extends into the similarly warm temperate area of the lower Yangtze River. This is the particular species believed by many to be the origin of the mythical Chinese dragon.

As we saw in Chapter 2, the fossil record shows us that crocodiles were considerably more diverse in the past. They had included both long-limbed, fully terrestrial species and permanently aquatic, swimming ones with the limbs reduced to flippers. At least one species was a land-living herbivore. However, only the amphibious kinds have survived until today. The body is very well adapted for this way of life, and they can quickly move between the land, where they spend much of the day basking, and the water in which they mostly feed and to where they flee if threatened. The long, powerful tail is used for swimming, while the legs though relatively short are still capable of quite fast running on land. Crocodilians are also the most interesting reptiles as far as behaviour is concerned. Unlike almost any others, they have a social structure involving various ways of communicating, and they exhibit parental care.

Despite the similarity amongst them, modern crocodiles are divided into three separate families. The Crocodylidae occur throughout the tropical world; the Alligatoridae are the alligators and caimans, all of which are American apart from the Chinese alligator. There are few ways to tell a crocodile (Figure 23b) from an alligator (Figure 23a) by their external appearance, except that the lower fourth tooth of a crocodile is visible even when the jaws are closed, the famous 'crocodile smile'. The third family, the Gavialidae, consists of just two gharial species. The true gharial is a large crocodilian, up to 6 m in length, whose habitat is fast-flowing, clear regions of the Ganges and other rivers of the northern Indian subcontinent. The smaller false gharial occurs in

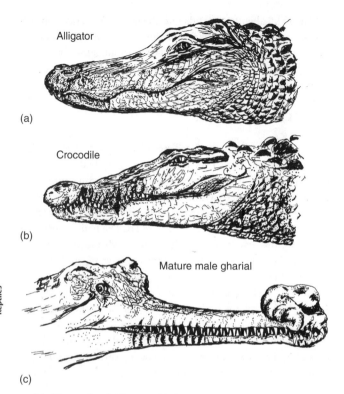

Alligator

(a)

Crocodile

(b)

Mature male gharial

(c)

23. (a) **alligator head;** (b) **crocodile head;** (c) **mature male gharial head.**

the Malaysian archipelago. Gharials are the most distinctive group of crocodilians, instantly recognizable by their extremely long, narrow snout. In the case of the male gharial, it ends in a curious bulbous knob, called the narial excrescence (Figure 23c).

Crocodile skin

Crocodiles are extremely well protected by their skin. As in all reptiles, the scales are small, thickened areas of what is otherwise the continuous keratin layer that makes up the epidermis. Narrow strips of softer, thinner keratin lie between the scales making the skin as a whole flexible enough to not restrict body movements too much. The squarish scales lying in rows over the back (Figure 24a) are particularly thick, and each one is stiffened by a ridge in the middle, making the animal even less vulnerable to a potential predator, such as one of the large cats. For more protection still, there is a layer of bony scales, the osteoderms, lying beneath the horny layer. In older individuals, the osteoderms often become exposed on the surface as the horny scales wear away. The flanks and the belly are softer, with fewer scales, and in many species there are no osteoderms in these regions. For this reason, these parts are favoured for crocodile leather goods such as shoes and handbags, a major cause of the decline of many species, at least until the advent of crocodile farming, and more effective conservation measures.

(a)

(b)

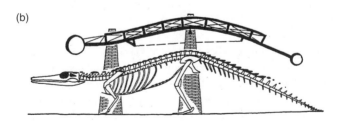

24. (a) Nile crocodile showing scale pattern; (b) crocodile vertebral column and dorsal scales modelled as a cantilevered beam.

The heavy osteoderms along the back of a crocodile are important for another purpose besides protective armour against enemies. Ligaments connect them with the vertebral column, which creates a stiff girder along the back to support the animal's body (Figure 24b). This makes it easier for the crocodile when walking rapidly on land to keep its body off the ground.

Although the general rule is that reptile skin is highly resistant to losing water, crocodile skin is far more permeable than that of strictly terrestrial reptiles. There is no evidence that this is a particular adaptation to do with controlling the animal's water balance. More likely it is simply a matter of saving on the production of waterproofing lipids in an animal living in a habitat where water is always freely available.

How crocodiles breathe: lungs and hearts

It came as a surprise at first when it was discovered that crocodiles have lungs and hearts that are more like those of birds than of other reptiles, even though no one ever doubted that birds and crocodiles are related to one another. Knowing this is important for two reasons. First, it helps us to understand why the physiology, activity, and behaviour of crocodiles are different in certain ways from other reptiles. Second, it opens a window into what the life and physiology of dinosaurs might have been like. Birds are technically small, living dinosaurs, so that whatever features crocodiles and birds share might reasonably be expected to have existed in the extinct dinosaurs too.

The lungs of other reptiles are flexible bags into which air simply enters and leaves by the same route, as they are inflated and deflated by movements of the ribs. It is called *tidal flow*. In crocodiles, as in birds, there is a *through-flow* system. The very fine tubes in the lung, where the oxygen and carbon dioxide pass between the air and the blood in the blood capillaries, are called parabronchi (Figure 25b). They are arranged so that the air always

(a)

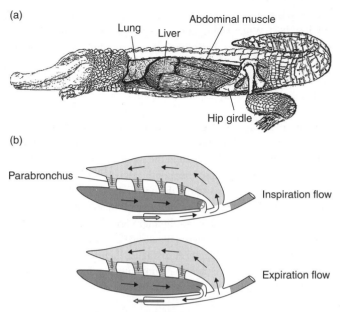

25. **Lungs and breathing mechanism of the crocodile: (a) lungs, liver, and abdominal muscles; (b) simplified diagram of the airflow through the lungs during inspiration (above) and expiration (below).**

flows in the same direction through them, during breathing in and breathing out. When breathing in, the air is drawn into the upper part of the lung. When breathing out, it flows through the parabronchi into the lower part of the lung and leaves the body from there. The advantage of this system is that more of the oxygen in the air is taken up by the blood, because the air in the parabronchi always flows in the opposite direction to the blood that is flowing in the capillary blood vessels closely alongside. This arrangement is technically called a *counter-current flow*. The way in which the lungs of crocodiles are inflated and deflated, the actual breathing mechanism, is unique to them. The back of the thoracic cavity that houses the lungs is connected to the liver, and the liver in turn is connected by muscles to the hip girdle (Figure 25a). When these muscles contract, the liver is pulled backwards expanding the thoracic cavity and lung, so drawing in air through the nostrils. Muscles of the abdomen then contract and squeeze the liver forwards, contracting the thoracic cavity, deflating the lungs, and expelling the air.

Another important feature of crocodile breathing is called the *secondary palate*, which is essential for the times when the animal is almost completely submerged in the water. A sheet of bone in the roof of the mouth separates the mouth cavity below from the nose cavity above. The air passes from the nostril directly to the epiglottis, which is a valve controlling the entrance to the lungs at the back of the throat. The crocodile can therefore breathe through its nostrils, which lie slightly raised up at the tip of the snout, even though the rest of the head is at or slightly below the water level. The epiglottis also prevents water from entering the lungs when the crocodile opens its mouth and grabs underwater prey.

The way in which the heart drives the blood around a crocodile's body is also partly, but not completely like a bird. Blood always needs to be directed to two places. It must go to the lungs to pick up oxygen and give up carbon dioxide. Then it must go to the rest of the body to supply the oxygen and remove the carbon dioxide from the animal's cells and tissues. The most efficient way to do this is called *double circulation*. In reptiles other than crocodiles, there is a flap of tissue inside the heart that directs the blood in the correct directions, but does not completely separate it into two streams. In crocodiles, the heart is completely divided into two sides by a sheet of tissue called a septum. Blood coming from the body tissues that has given up its oxygen flows into one side of the heart, and from there it is sent to the lungs. It returns from the lungs now loaded with oxygen and enters the other side of the heart. From here it is directed back to the body tissues. So far, crocodile and bird hearts are basically similar, but there is one big difference. Crocodiles have a small opening in the main vessels as they leave the front of the heart, given the wonderful name of the *formen of Panizza* after the Italian who discovered it. Normally it is kept closed, but when it opens, it allows some of the blood on the way to the lungs to be diverted back to the body instead. This looks at first sight like a sort of leakage that must reduce the efficiency of gas exchange, until we remember that at times when the crocodile is completely under water and holding its breath, it would be a waste of energy to pump all the blood to the lungs.

There is something that does not quite make sense about the action of the lungs and the heart of crocodiles. The unidirectional flow of air through the parabronchi of the lungs and the double circulation of the heart are both ways of increasing the amount of oxygen available to the body. They are essential for animals with a continuously high metabolic rate, such as the birds. But, as we described earlier for reptiles in general, crocodiles are *ectothermic*, with the same characteristically low metabolic rate and level of aerobic activity as others. The answer to this puzzle may be that modern crocodiles are descended from ancestors that had higher metabolic rates. If this is true, it has very interesting implications. One is that the commonly held view that the ectothermic reptiles are somehow always inferior to the endothermic birds and mammals is misguided. The second is that archosaurs other than crocodiles, including all the dinosaurs, must have had relatively high metabolic rates and activity.

How crocodiles move

There are no burrowing, tree-climbing, or gliding crocodilians, and since the end of the Mesozoic none specifically adapted for either fast running on land, or permanently swimming in the sea. But compared to this absence of adaptations for specialized ways of moving, the locomotion of an individual crocodile is remarkably versatile. It has a variety of different styles of locomotion for different circumstances during the course of its life. Only a few species have ever been fully investigated from this point of view, mainly the American alligator and the Nile crocodile, but given the similarity of the body form of all crocodilians, there is little doubt they all show much the same versatility. The key to this lies in some unique features of the skeleton and muscles. One is the coupling by ligaments of the osteoderms along the back with the underlying vertebrae and ribs, to create a strong girder between the front and hind legs that can support the weight of the body off the ground (Figure 24b). Another is a special extra joint in the ankle that increases the possible range of movements of the foot. A third locomotory adaptation is the deep socket in the hip bone that extends and controls the movement of the thigh bone, and therefore of the hind leg as a whole.

When on land, a crocodile that is simply basking in the sun to warm up lies on its belly with its legs stretched out alongside (Figure 26a), and if it needs to adjust its position, or go into the water to cool off, it uses a simple lateral undulation of the body and a sprawling gait of the legs, called the *belly crawl*. If, on the other hand, it is travelling a longer distance, or is disturbed by the approach of a possible predator, it turns its legs underneath its body, raises itself completely off the ground, and moves by the *high walk* (Figure 26b). This is faster than the belly walk. For even higher speeds, such as escaping a serious immediate danger, a rather odd sort of *bounding gallop* may sometimes be adopted, especially by young crocodiles (Figure 26c). The back legs act together to drive the body forwards and clean off the ground, and then they stretch forwards in front of the front legs, ready for the next bound. By this method, a Nile crocodile has been reported to reach a speed of over 40 kph, enough to briefly outrun most predators and reach the safety of water, but so energy consuming that it can only be kept up for a very short time. There are also records of crocodiles raising themselves up onto their hind legs and briefly adopting a bipedal gait, when chasing prey.

In the water there are also several different styles of locomotion. While at rest, a crocodile typically floats, with the body held at an

26. Modes of crocodile locomotion: (a) the belly crawl; (b) the high walk; (c) the gallop; (d) the torpedo run; (e) nose clamp capture of an antelope.

angle and only the nostrils, eyes, and ears above the surface. When it is slowly moving around, the body is held horizontally and straight, the legs are tucked into the sides, and gentle undulations of the tail propel the animal.

Another form of slow moving is called *vertical patrolling*, in which the body is held vertically, the limbs gently paddle, and only the head is above the surface. At other times, such as when chasing a fish or about to launch an attack onto the bank to snatch ts prey, a crocodile will indulge in *torpedo swimming* (Figure 26d). It accelerates rapidly with strong beats of its tail, and then becomes perfectly straight as it shoots itself forwards towards its target.

How crocodiles feed

All crocodilians are predators, though they often include carrion in their diet. Their food varies from insects up to large mammals such as antelopes, but especially fish, amphibians, and birds. The normal way of catching prey on the land or the margin of the water is to wait hidden in ambush for a victim and pounce quickly upon it (Figure 26e), because crocodiles do not have enough energy for more than the briefest of chases. The simple but sharp teeth set in long jaws are typical of fish eaters. They are eminently suitable for grasping and holding on to this slippery prey, but they are incapable of chewing it. Instead, if the victim is fairly small such as a fish or a frog, it is swallowed whole by the simple method of the crocodile holding up its head and letting the prey slip back down its throat under gravity. Part of the stomach is a thick-walled, muscular bag that contains hard pebbles called *gastoliths*, which the crocodile has deliberately swallowed. Once here, the food is ground up into smaller pieces before continuing back into the intestine for digestion.

The gharials are the most committed fish eaters of the modern crocodiles. They live almost totally on fish and swimming frogs, which they catch by slashing their narrow snout sideways towards the prey. In fact, gharials are the most aquatically adapted crocodilians in general. Their legs are relatively small, and they rarely venture out onto land except for the female to lay her eggs. Virtually all other crocodile species have a wide, unselective diet. Fish is usually the main component, typically being captured by the front of a crocodile's jaws after a fast underwater torpedo run towards it. A Nile crocodile has sometimes been seen encircling a shoal of fish with its body to trap them, before turning inwards and grabbing one. But virtually any underwater prey in addition to fish will be caught, including aquatic crustaceans, insects, amphibians, and small freshwater turtles.

Catching larger prey on land is more complicated because of the crocodile's inability to kill directly. They show considerable versatility, but the key element is surprise, upon which a successful capture invariably depends. The crocodile will remain more or less invisible and unmoving in the shallow water, or lying on the bank. If an unwary bird or mammal comes within range, the attacker makes a rapid lunge and attempts to grasp the victim in its jaws. If successful, it promptly retreats into the water dragging its prey with it, and holding it under the surface until it drowns. Larger prey such as an antelope, or even an animal the size of a zebra or giraffe in the case of the Nile crocodile, may be similarly ambushed, grasped by the nose or neck and dragged under. Another common method used is to rapidly emerge and knock the prey into the water by a powerful sideswipe of the tail. It is then grabbed and pulled down. Prey that is too large to swallow whole, and too tough to tear up immediately, is kept under water, wedged between submerged branches or rocks for a few days, until enough decomposition has taken place to allow it to be pulled to pieces. The crocodile will then grasp a part of the prey and by spinning its whole body round and round in the water, tear off a chunk. This is often a communal activity, and I have watched a crowd of some twenty Nile crocodiles spending several days pulling a drowned giraffe to pieces in this way.

Around seven crocodilian species are known to have attacked humans, although the Nile and saltwater crocodiles, and the Florida alligator, have the worst reputation for aggressiveness, showing little hesitation in attacking a person if the opportunity arises. Particularly at risk are young children playing by the waterside. Most encounters with crocodiles result at most in serious injury, such as loss of a limb. But deaths do number a few hundred each year worldwide, sometimes catastrophically following a boatload of people capsizing.

Crocodile sense organs

As amphibious animals, crocodiles need to have sense organs suitable for use in two different media, air and water. They must also have sense organs for use both in very bright sunlit conditions and in darkness underwater or at night. Vision is their most important sense for catching food, and their eyes are amazingly versatile. To start with, crocodiles have good colour vision, as in lizards. There are three kinds of colour sensitive receptors in the retina, cones sensitive to violet, green, and red light respectively. When used in the air, crocodile eyes are very acute, having about the same ability to see detail as a cat. This is thanks to a small region of the retina of the eye called the *fovea* or yellow spot, which has a very high density of light-sensitive cells. In the cat, as in ourselves, the fovea is a small circular area, and we have to keep moving our head to maintain the focus of a moving object on it. The crocodile's fovea is different; it is a horizontal band across the middle of the retina. The eyes are positioned slightly above the level of the top of the head, and so they are exposed to the air as the animal floats near the surface of the water. The whole of the horizon is in focus on the fovea at the same time. A potential prey is continually watched as it approaches the water to drink, but without the slightest movement of the virtually invisible crocodile disturbing it, until it is close enough to be attacked.

Due to the different light-transmitting nature of air and water that we are so aware of when we go swimming, eyes adapted for working on land cannot focus very well under water. But crocodiles have solved this problem by a special transparent membrane called the *nictitating membrane*. When the animal is submerged, the membrane is drawn across the front of the eye and acts like swimming goggles, significantly improving the vision at the same time as protecting the surface of the eye.

The eyes are also wonderfully adapted for the different light levels of night and day. In the first place, the pupil of the eye can vary in shape. It is fully round in the dark to allow as much light in as possible, but reduced to a narrow slit in bright sunlight, restricting the amount of light entering. Secondly, there is a layer of reflecting crystals behind the retina called the *tapetum*. In low light, such as at night, the crystals increase the amount of light reaching the light-sensitive cells of the retina by reflecting it back. This is why torchlight makes the eyes of a crocodile shine brightly in the dark. Many night-time animals have a similar crystalline tapetum, but the crocodiles have gone further, and have a layer of pigment cells in front of the crystal layer. During the daytime, the pigment cells expand and act like sunglasses by preventing the reflection of light back to the retina. In the dark, however, the pigment cells contract and allow the tapetum to do its job.

The next important sense organs are found only in crocodiles. They are little dome-like clusters of cells called ISOs, which stands for *integumentary sense organs*. They lie on the scaly surface of the head, and in some species on body scales as well, and are very unusual because each ISO is sensitive to three different kinds of stimuli. They are extremely sensitive to the surface pressure waves in the water, created for example by potential prey swimming at or near the surface. This makes them important for hunting at night. The ISOs are also very sensitive to temperature, which makes them important in behaviour regulating body temperature, such as basking to warm up or returning to water to cool down. Thirdly, the ISOs are sensitive to the acidity of the water. It is not so obvious why this information is important; probably it is used to indicate excessively stagnant water full of rotting vegetation, or a decaying corpse.

Hearing in crocodiles is by means of a normal ear drum and, while not particularly sensitive, is important for sound communication between individuals and between juveniles and parents. It is of little use underwater though, because the ears are closed off by muscles to prevent water from entering. The crocodile sense of smell is quite good. It is helped by pumping the floor of the mouth up and down to draw air into the nose cavity, and by this means a crocodile can detect meat and carrion from a long way away. It is also important when submerged. The mouth can be opened and the head moved from side to side to detect such food. As in most reptiles, olfaction is a part of their social communication. Signalling molecules, the pheromones, are produced in glands in the throat and alongside the cloaca and are important in mating and nesting behaviour.

Social and reproductive behaviour

The social behaviour of crocodiles, especially courtship and the care of the young, is much more elaborate than in any other reptile group. The details differ from species to species, but in general the males are territorial, there is elaborate courtship behaviour, the eggs are laid in a prepared nest guarded by the female (Figure 27), and the hatchlings are actively cared for at first by the mother.

The Nile crocodile is one of the species whose behaviour has been most studied. Outside the breeding season, they often live in mixed groups, basking together on the bank and sharing a large kill. The breeding season normally starts a little before or early on in the rainy season, when the males start to acquire territories to

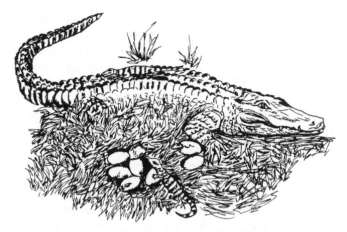

27. Female Nile crocodile caring for eggs and hatchlings.

protect their own nesting sites, sources of food, and females. Rival males are fought off by sparring with the jaws, and by head to head contact. Courtship by the male in the presence of a female is a drawn-out process lasting several hours, during which he opens his mouth, bellows, slaps the surface of the water with head and tail, and produces a strong scent from glands alongside the cloaca. The female acknowledges the male's presence with low-pitched sounds of her own, and either flees or eventually allows the male to mate with her.

Once successfully mated, the female uses her front legs to create a nest by digging a hole in a sandy beach 10–20 m from the water's edge. Here she deposits around forty hard-shelled eggs and covers them over with vegetation. Like other reptiles, the sex of the offspring depends on the temperature at a critical period during development. Males are produced at 32–3°C, females at temperatures above and below this. We do not know to what extent the female selects the nesting site carefully in order to determine the sex of her offspring. However, she lays her eggs in layers, and therefore different ones must experience different temperatures. Perhaps this is sufficient to guarantee a mix of males and females. Once egg-laying is complete, the female remains close by to protect the eggs from being stolen. Nile monitor lizards are the main predators of crocodile eggs, but several mammals including mongooses and hyenas are also threats.

After about three months, the eggs start to hatch. The emerging juveniles give off high-pitched calls that alert the mother, who comes and opens the nest, and with great delicacy uses her jaws to crack open any unhatched eggs. Once all the eggs have hatched, she ferries the young in her mouth to the water where, for the next two years, she stays with them and continues to protect them, and communicate with them using distinct maternal calls. For the following few years, while they are still small, the growing juveniles tend to hide away from larger crocodiles because cannibalism is certainly not uncommon. Finally, after about 8–12 years, the offspring have grown to around 2.5 m long, become sexually mature, and are ready to breed.

Chapter 6
Chelonians

Chelonians are the tortoises, terrapins, and turtles, and are amongst the most bizarre vertebrate animals living today. These three common names are confusing because they do not correspond to groups of chelonians recognized by scientists. The name 'tortoise', strictly speaking, means the fully terrestrial, dome-shelled members forming the family Testudinidae. The freshwater chelonians belonging to the family Emydidae are called 'terrapins' in Europe, but Americans refer to them as 'freshwater turtles'. The name 'turtle' used on its own sometimes refers to chelonians in general, and sometimes just to the marine chelonians. It is also used as part of the common name of other families, such as mud turtles, soft-shell turtles, side-necked turtles, and even the American box turtles which look very like true tortoises. Worse still, Australians, who do not have any native testudinids, nevertheless call all their land and freshwater chelonians 'tortoises'!

The single most characteristic feature of chelonians is, of course, the shell. In place of the flexible layer of small scales that other reptiles use to protect the body, they are covered by a small number of very large scales, immovably joined to one another. Inside the shell (Figure 28a and b), the number of vertebrae has been greatly reduced, and all traces of the lateral undulation of the body typical of most reptiles have vanished. As we shall see, this

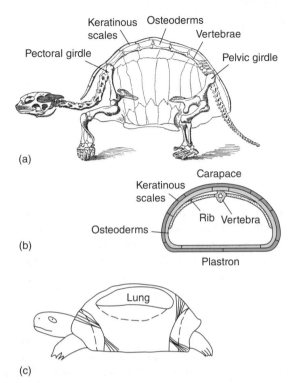

28. The chelonian body plan: (a) the skeleton of a tortoise seen with the left side of the shell removed; (b) transverse section through the shell, vertebral column, and ribs; (c) breathing mechanism, showing the membranes bounding the floor of the lung and the front and back boundaries of the body, and the muscles that move them. Dashed lines show their position after inspiration.

greatly affects other aspects of their anatomy and biology, such as how they walk, breathe, and feed.

The roughly 320 species of living chelonians fall into two groups. One of these is called the Pleurodira or the side-necked turtles, given this name because they withdraw their head into the shell for protection by a sideways bending of the neck. Pleurodires are found only in freshwater habitats of the southern hemisphere in South America, Australia, and Africa. All the remaining chelonians belong in the Cryptodira, which are by far the most diverse of the two groups. They withdraw their heads by a vertical bending of the neck, such as we see in a typical tortoise, although some species, notably the sea turtles, have lost the ability to withdraw the head at all.

The chelonian body plan

The whole structure of the body of a chelonian is based around its remarkable shell (Figure 28a and b). This is made up of the top part called the *carapace*, which may be anything from almost flat to highly domed. It is connected between the front and the hind legs to the bottom part, called the *plastron*, which covers and protects the underside of the body. The shell consists of two layers. The outer layer that we see in the live animal is made up of large, horny scales, or scutes as they are called (Figure 32a and b). There is a row of typically five down the centre of the carapace called neural scutes, and a ring of smaller ones around the edge called marginal scutes. The space in between is filled in with just four pleural scutes on each side. The plastron is made up of six pairs of large scutes. There is then the inner layer of the shell, composed of large, bony plates called osteoderms. Eight neural osteoderms along the top of the carapace are immovably fused to the tops of the eight trunk vertebrae. On each side of these there are eight costal osteoderms fused to the broad ribs. By this arrangement, the shell as a whole is rigidly attached to the skeleton.

This double layered shell creates an extremely strong structure for protecting the body, and the relative invulnerability of most chelonians is increased by being able to withdraw the head, limbs, and tail inside the shell. Despite this, small chelonian species, and juveniles of larger ones, are prey to many animals. Some, such as monitor lizards and crocodiles, simply swallow freshwater turtles whole. Other predators are able to crack open the shell of even quite big individuals. The southern ground hornbill of Africa, for example, has a powerful enough beak to hammer the shell open directly, while amongst the mammals whose jaws and teeth are strong enough are otters and the large cats. Bearded vultures and several other birds of prey have learned to catch tortoises, fly up, and drop them onto rocks below to break them open.

The shell can also protect the animal from excessive loss of water by evaporation. In the land tortoises that live in arid and desert regions, the outer layer of horny scutes is very well waterproofed, and withdrawing the extremities completely into the shell reduces even more the exposure of the body to the dry air. The shell also helps to insulate the body against overheating, and many tropical tortoises can live in open or light scrubby country, with little of the shade that lizards and snakes need in the middle of the day.

Evolving such a useful thing as a rigid, protective shell had a profound effect on other aspects of chelonian life. The very short, inflexible body prevents lateral undulation from playing any role in walking or swimming. Add to this the fact that chelonians on land have legs that are short enough to be withdrawn into the shell, and stout enough to carry the increased weight of the animal, and it is clear that the way the animal moves needed to evolve considerable modifications. The limb girdles that support the limbs (Figure 28a) lie inside the shell, and are strongly attached to it to transmit the walking forces produced by the legs. The position of the shoulder girdle was for a long time very puzzling because it lies *inside* the rib cage, compared to all other reptiles, where it lies outside. What happened was a change in how the ribs develop in the embryo. In other reptiles, the ribs grow downwards inside the shoulder girdle to surround the body and create a rib cage. In chelonians, the ribs stay up above the girdle and fuse to the costal plates of the carapace.

The short, stout limbs project outwards through narrow gaps in front of and behind the bridge where the carapace meets the plastron. In the land-living tortoises, they end in compact feet like miniature elephant's feet, but with robust claws. When walking, the length of each stride is small, and given the weight they must carry, the limbs can only move slowly and deliberately. The reputation that tortoises have for slow moving is certainly justified compared to other reptiles. Giant tortoises have a top speed as low as 0.3 kph, although smaller species can sometimes reach 8 kph or more, which is about the same as a fast human walk. However, the usual reaction of a tortoise to a threat is not to flee, but simply to freeze and withdraw head, limbs, and tail into the shell. Running as fast as they can is more often for chasing a mate than for escape. We shall meet later the ways in which some chelonians overcame the limitations on their locomotion by adapting to life in water.

Chelonians lack a flexible rib cage, and therefore another consequence of the evolution of the shell is that they can no longer fill their lungs by expanding the thoracic cavity as other reptiles do. A completely new method of breathing evolved (Figure 28c). The large lung is attached above to the inner surface of the carapace, and below to the flexible roof of the main body cavity. Flexible muscular membranes close off the front and back openings between the carapace and the plastron. Special muscles, and also a rocking movement of the shoulder girdle, draw these membranes inwards and outwards. This alters the pressure of the body cavity and in turn the lung, so that air is drawn in and out through the nostrils.

The locomotion of chelonians is relatively slow and ponderous, which leads to a third biological consequence of the evolution of the shell. They can no longer be active, agile hunters as their amniote ancestors had been, chasing and catching live insects and small vertebrates for food. Instead, most have come to rely on a diet of plants and slow-moving worms and snails. This is much more easily gathered food, but it tends to be tough, and in place of snapping jaws, chelonians have had to evolve a slower, but stronger bite. The muscles for closing the jaws are large, and the jaws themselves are short, which increases the bite force. Even more unusually, no living chelonian has so much as a single tooth, for they have been replaced by upper and lower horny plates made of tough keratin, like a bird's beak. Plant material in particular is often abrasive, and teeth would quickly wear down, but tooth plates have the advantage that they can keep on growing, to keep pace with the wear. In the more strictly herbivorous species such as tortoises, the tooth plates are flat and heavily ridged for grinding. More carnivorous chelonians, such as the freshwater groups, tend to have a sharp, serrated edge to the tooth plates for cutting up slow-moving molluscs, worms, crustaceans, and small fish.

Despite the limitations that the shell imposes, there are many variations upon the general body plan and way of life amongst chelonians. There may not be any fast-running hunters or tree climbers amongst them, but there is certainly an impressive range of land, freshwater, and marine kinds that we will look at next.

Life on dry land: tortoises and box turtles

Tortoises are the members of the large family Testudinidae, which live on land and typically have a high domed shell of a special three-dimensional shape called a Gomboc. It is flat on one side and convex on the other, and has the remarkable property that it always tends to roll over onto the flat side, whatever position it is placed in. Tortoises evolved the Gomboc shape long before two Hungarian mathematicians discovered it and showed that the shell of the Indian star tortoise (Figure 29) is an example. The importance for these animals of anything that assists them to retain their proper orientation on four legs after accidentally rolling over can hardly be exaggerated.

There are about fifty species of tortoises. They are found in all the tropical and subtropical parts of the world, with the notable

29. The Indian star tortoise.

exception of Australia, but are most diverse in southern Africa. In size they vary from giant tortoises weighing over 300 kg, to the 6–8 cm long, 100 g male speckled padloper, which lives amongst the rocks in semiarid regions of Namaqualand, feeding on small succulent plants. Indeed, all tortoises are primarily herbivores, although several are omnivores that will also consume any other food they come across, such as worms and molluscs, and even carrion. They are especially well adapted for arid habitats. The high domed shell reduces the animal's surface area, which helps it to stay cool in hot conditions by reducing the amount of heat absorbed, and to reduce the loss of water. Resistance to water loss is increased even more by withdrawing the head and tail completely into the confines of the shell, and closing off the front and back openings between the carapace and the plastron using the heavily scaled legs. The bladder is large enough to store fresh water, when available, to provide the tortoise's needs for several weeks.

The shell serves other roles too, including camouflage. Several species, like the star tortoises of India (Figure 29) and the

Kalahari tent tortoise of Africa, have an attractive pattern of bright yellow radiating stars on each of the carapace scutes, which resemble the dappled shadow effect of low vegetation. The shells of males may also be used as weapons, for example the rare ploughshare tortoise of Madagascar, in which rival males attempt to turn one another over by means of a broad blade that projects forwards from the front edge of the plastron.

The giant species of Aldabra in the Seychelles, 1,500 km off the east African coast, and the Galapagos Islands 900 km off the coast of Ecuador are certainly the most spectacular tortoises. In 1835, during his stay on the Galapagos Islands, Charles Darwin was struck not only by the giant size of these extraordinary reptiles, but also by reports of how the particular island a specimen came from can be identified by the shape of the shell. It is one of the cases of natural geographic variation he so astutely noted that eventually contributed to his theory of evolution by natural selection. In the past few centuries, similarly large tortoises lived on several other Indian Ocean islands, and on parts of the mainland. Evidently they can be relatively easily transported over long distances by sea aided by the large size, and the high position of the lungs which makes them float in an upright posture. Their ability to survive long periods without food and water is no doubt also important. The largest specimen ever recorded was an Aldabran tortoise in London Zoo that weighed about a third of a tonne, and whose shell measured 1.4 m. All tortoises are exceptionally long-lived, especially these giant species. There are claims that some individuals were over 200 years old, although the longest undisputed record holder was a specimen of a Galapagos tortoise in the Australian Zoo, which lived for 175 years.

Several kinds of tortoises can burrow, but the real experts are the gopher tortoises of hot and arid regions of North America.

30. The pancake tortoise.

Using their spade-like front legs, they can dig a burrow that is up to 30 m long and 3 m deep, where they avoid the excessive heat. Adaptation to another way of life involving hiding is found in the oddest, most specialized tortoise of all. The shell of the pancake tortoise of eastern Africa (Figure 30) is flat, and the osteoderms so reduced that the carapace and the plastron are both flexible. This allows it to live and move amongst narrow crevices between rocks, and when threatened, it tightly wedges itself by filling its lungs with air to expand the body, making it virtually impossible to dislodge.

Another group of chelonians are very similar to tortoises, but have independently evolved a domed shell, short and stout walking legs, and a life spent largely on land. These are the box turtles of North America, which are members of the otherwise mainly freshwater family Emydidae. The eastern box turtle lives in open woodlands from Florida right up into Canada, eating a mixed diet of plants and invertebrates. Like a few of the true tortoises, box tortoises have a hinge across the plastron so that the front and the back sections of it can be drawn upwards to partly close off the gaps at the front and rear end and so help protect the body.

Life in ponds and rivers: side-necked turtles, terrapins, and others

Adapting to swim and feed in water was evidently an easy way to modify the chelonian body plan, because members of several groups have independently done just that. The shell is much flatter, which creates a streamlined shape that can more easily move through the water. The limbs, though short and stubby, have webbing between the fingers and toes and work quite well as paddles for swimming. Indeed, so well suited are turtles for living in water that some zoologists believe the chelonians as a whole evolved originally as aquatic organisms, and that the dome-shelled terrestrial tortoises and box turtles are the more specialized.

Pleurodires, the side-necked turtles of the southern continents, are all freshwater, or at least marsh dwellers. They are also known as snake-necked turtles because of the long, flexible neck, which is used when seeking food. For most species, this consists of slow-moving invertebrates such as worms and molluscs, but crustaceans and small fish will often be grabbed as they pass by. The neck is able to bend horizontally and lie sideways within the gap between the carapace and the plastron, with one side of the head facing out from its protective shield.

The most extraordinary pleurodire is the matamata (Figure 31), a large, fierce predator that dwells in shallow, slow-moving streams and marshes of South America. Its carapace is quite small and rough and there are three long ridges down it. The neck is long, the head flat, and both of these are fringed with ragged flaps of skin and tubercles. The overall effect is that the body looks just like a pile of dead leaves and bark, and is almost invisible as it remains motionless at the bottom of the murky water. The front of the head extends into a narrow snorkel that is used to breathe air from the surface with scarcely any movement of the animal. When a fish passes close enough, it is detected by the sensitive

31. The matamata.

skin flaps, whereupon the matamata rapidly opens its large mouth, sucks the fish in, and swallows it whole. One unexpected consequence of the unattractive appearance of matamatas is that they are spurned as food by local people, even though other species of chelonians are readily caught and eaten.

Several different groups of cryptodires, the other branch of chelonians, are adapted to life in fresh water. All of them have the same general shell and limb anatomy, and most feed on a mixed diet that includes invertebrates and often fish, as well as vegetation. The North American pond slider (Figure 32a) is a typical example of an emydid, with its low, rounded shell and webbed feet. The European terrapin is similar. It is a modest-sized species, up to about 30 cm long, and dull yellow-brown in colour, and its range extends through much of Europe, as far north as Poland and the Baltic States. It lives in slow-moving waters overhung with vegetation, feeding on crustaceans, worms, and molluscs. Like all chelonians, terrapins must lay their eggs on land. The female carefully selects a suitable nest site within 100 m of the nearest water, lays her eggs, and then, in common with all chelonians, takes no further interest in her offspring.

The ferocious-looking snapping turtles of America are amongst the largest freshwater turtles in the world. The alligator snapping turtle of the south-eastern states of the USA can reach a body

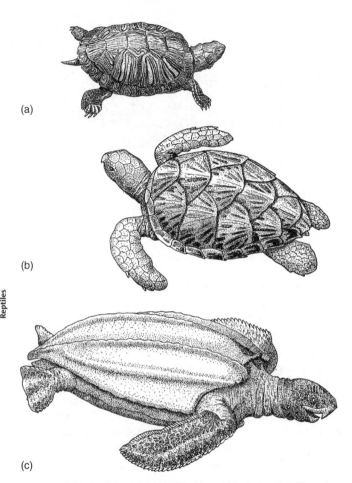

32. Aquatic chelonians: (a) pond slider (emydid); (b) hawksbill turtle; (c) leatherback turtle.

weight over 100 kg, and shell length over 80 cm. It spends most of its time at the bottom of shallow water, swimming little but feeding on almost any animal, vertebrate or invertebrate, passing close enough to be grabbed by the large head and powerful jaws.

Fish are caught with the help of an attractive, fleshy lure on the end of the tongue that looks like a wriggling worm.

The strange soft-shelled turtles are the chelonians most highly adapted of all for freshwater life. They include the largest of all freshwater species, the extremely rare Swinhoe's soft-shelled turtle of eastern China and Vietnam, which has a carapace over 1 m in length and a weight of 140 kg. As their name indicates, the flat shell has lost all the horny scutes, and all the osteoderms apart from a row down the centre of the back. In their place, the back has a soft, leathery covering. Soft-shelled turtles are good swimmers, and spend most of the time under water. This is helped by an ability to breathe oxygen from the water using the skin, the walls of the cloaca, and the mouth cavity, all of which are very well supplied with blood capillaries. The long neck can also be used as a snorkel to breathe air at the surface, especially in shallow, stagnant waters containing little oxygen. The head is an unusual shape. It is at the end of a long, flexible neck and there is a fleshy extension, a proboscis, at the front. Instead of the normal chelonian horny tooth plates, fleshy lips line the jaws. Feeding is by ambushing passing prey such as fish, as the turtle lies perfectly still at the bottom of the water, often buried in the mud.

Life in the sea: marine turtles

Chelonians could never be totally marine animals because, unlike the sea snakes and the ichthyosaurs of the Mesozoic, for some reason they have never been able to lose the egg-laying habit and bear their young live, directly into the water. Therefore, at the very least, the females would always have to emerge onto land to create nests and lay their eggs. Nevertheless, the living species of marine turtles get as close to a permanently marine life as otherwise possible, with a variety of morphological, physiological, and behavioural adaptations; the males rarely if ever come ashore once they have entered the sea as hatchlings, and the females only emerge briefly and laboriously during the egg-laying season.

There are two groups of marine turtles. Six species, including the green, hawksbill (Figure 32b), and Ridley turtles, make up the family Cheloniidae, while the leatherback turtle (Figure 32c) is the single living member of the related family Dermochelyidae. The latter differs from the others mainly in the nature of the shell, which consists only of a tough, leathery skin in which are embedded thousands of tiny bony platelets. The carapace has seven longitudinal ridges from front to back, which act like fins to stabilize the body when swimming. Living in the sea is helped by large body size, which reduces heat loss, but even so, most sea turtles remain within tropical and warm temperate areas of the world's oceans. The smallest of them, the Ridley turtles, are about 75 cm long, and they are restricted to warmer waters. The largest is the leatherback, whose carapace reaches close to 2 m in length and is the only marine turtle that can survive occasional forays into higher latitudes, north of Scandinavia and south of Australia.

The sea turtle carapace is flattened to reduce the resistance to movement in the water. There are no gaps between the carapace and plastron for the head and legs to be withdrawn into, which improves the streamlined shape. The way they swim is different from the freshwater turtles, which use all of their four limbs as paddles pushing back against the water. The front legs of sea turtles are longer, more slender, and pointed, with no outward sign of separate fingers. In principle they work like a bird's wings, beating up and down to create the forwards force. Unlike a bird which has to contend with gravity all the time, turtles are suspended by the buoyancy of the water and so the upbeat as well as the downbeat of the flippers drives the animal forwards. The back legs are shorter, and used for steering and manoeuvring. The maximum swimming speed of turtles is hard to measure. A leatherback turtle has been recorded at over 30 kph when frightened, although around 10 kph is a more usual cruising speed for this and other species. Breathing in sea turtles is similar to other turtles, using movements of the limb girdles to enlarge and decrease the volume of the body cavity, but is more efficient because they have larger lungs. In the leatherback turtle, breathing is helped by the loss of the rigid shell, which allows the whole body to expand and contract to some extent. The blood of sea turtles has a very high concentration of the oxygen-carrying molecule haemoglobin, and their body tissues can withstand a particularly high level of the waste product, lactic acid, that starts to build up once all the oxygen has been used up. Armed with these locomotory and breathing adaptations, sea turtles have a remarkable diving ability. Leatherback turtles can reach depths of over 1,000 m, and even the smaller species such as the hawksbill and Ridley turtles regularly dive to almost 100 m, and can stay submerged for over two hours.

The swimming ability of the marine turtles, however, is not good enough for them to be able to survive by catching fish, although Ridley's turtles in particular are sometimes found with catfish remains in the intestines. Otherwise, they are all feeders on either plants or slow-moving invertebrates. The green turtles prefer grazing on sea grass, one of the few kinds of flowering plants that grow submerged in shallow tropical sea water. The hawksbill turtle, with its prominent, sharp-edged cutting tooth plates along the edges of the mouth, lives mainly amongst coral reefs feeding on the abundant sponges and soft invertebrates there. Many sponges contain highly toxic substances to which the hawksbill itself is immune, but the flesh of these turtles may become highly poisonous and is frequently fatal to humans who eat it. The leatherback turtle is the most specialized feeder. It has no tooth plates in its jaws, and eats virtually nothing but jellyfish, which it often captures at considerable depths.

The most astonishing facet of the marine turtles' life is their life cycle. It involves the migrations the juveniles make, from the beach where they hatched out to the feeding grounds, and years later as adults back to exactly the same beach they had started from. It may be a relatively short distance, such as those undertaken by the populations of green turtles that both feed and nest on the shores of the Red Sea. For others it is a huge journey of several thousands of kilometres, such as is undertaken by the population of the same species that feeds in New Caledonia in the middle of the Pacific, and returns to the Great Barrier Reef off eastern Australia to breed. Loggerhead turtles hatched on Japanese beaches feed 10,000 km away off the coast of Mexico.

The story starts as the hatchling turtles use a special egg-tooth to escape from the eggs left buried in the sand nest and abandoned by the mother several weeks earlier. Acting together, they dig themselves out, and once they have fought through to the surface, they find themselves high up on the beach facing the most hazardous journey of their lives. They need to get to the relative safety of the sea, while trying to avoid the waiting flocks of predatory birds such as frigate birds and herons in the air, and the large crabs on the beach. Scrabbling across the sand at night, drawn towards the lighter sky over the sea, a large number never make it. Even those that do get this far now face an array of fish predators. Once in the water, they swim as quickly as they can towards the open sea, and experiments show that they are guided at first by the direction the waves come from. From this point, the tiny juveniles start their migration to the adult feeding grounds, living on plankton as they go. They are carried by the great circulating ocean currents called gyres, but the main sense they use for navigation is their ability to detect the magnetic field of the Earth. Amazingly, they can sense not only the direction of the magnetic lines, but also the angle of inclination relative to the horizontal. These two values combined give a unique reading for practically every point on the Earth's surface. This phase of the migration can take several years, but eventually the now much larger juveniles head away from the open ocean towards their particular destination. This is the shallower feeding areas off the continental coasts and islands where their parents had come from. Here they live and grow to adulthood over the next several years, before commencing the return journey. The Ridley's turtles breed

Reptiles

annually, but the others only migrate back to their original homes every few years. Whatever environmental, hormonal, or nutritional factors trigger them, the adults set off back to the very beaches where they had hatched so many years earlier. Again the Earth's magnetic field offers the navigational guide, although once close to the final destination, more local cues such as smell probably allow them to home in to the exact same spot.

Mating is a somewhat aggressive process on the part of the males, involving biting the female's legs and neck and clinging on top of her. This happens offshore, and after copulation the female stores the sperm, often from a number of males, so she can lay her eggs in batches over the course of several weeks. When ready to lay, she uses her flippers to drag herself up the beach to find a suitable place to dig a nest. The site must be safely above high water mark so the eggs will not drown, a line often indicated by the start of sand dune vegetation. But the further from the sea's edge, the greater the hazard will be for the hatchlings, and the sand itself has to be loose enough for them to dig themselves out. Once selected, the mother digs a hole about half a metre deep using her hind legs as shovels. Into this she lays about a hundred eggs and covers them over with sand to conceal them. She then departs back into the sea until ready to return, a week or two later, to dig another nest close by the first, and lay her next clutch. Finally, her last clutch having been deposited, the female returns to sea to follow the males, who have already departed on their epic return journey all the way back to the feeding grounds.

No one knows for certain why marine turtles undertake these often prodigious journeys. Having sandy nest sites on land away from the offshore reef and oceanic feeding sites is inevitable, as long as egg-laying is how they reproduce. But why the adults apparently never discover breeding sites closer to where they spend the majority of their life living and feeding is a mystery. Perhaps it is because the place where they themselves successfully hatched as juveniles has already proved to be suitable, and therefore offers a better guarantee of success for the next generation than would a new, untried place. Possibly, though, it is just that the navigational map inbuilt into the brain over millions of years is virtually impossible to modify. Whatever the truth, sea turtle migration is one of the great wonders of biology.

Chapter 7
The future of the world's reptiles

The periods of time that geologists divide the history of the Earth into are in most cases separated by brief episodes of severe worldwide extinction. Many of the existing species, occasionally a very substantial majority, disappear from the fossil record. In Chapter 2, we met the mass extinction that marked the end of the Permian Period, when over 90 per cent of species disappeared, and also the one at the close of the Mesozoic that witnessed the end of the dinosaurs and many other kinds of organisms. Even within each period, species continually become extinct at a low rate, only to be replaced by newly evolved ones. Extinction was not only the norm in the past, but just as essential a part in evolution as the origin of new kinds of organisms. The old must give way to the new. Exactly what causes a particular species to go extinct is never easy to say. Ultimately, it is due to some change in the environment that prevents the population from breeding so successfully: if its birth rate drops below its death rate for more than a short while, it is inevitably doomed. There are endless possible ways in which the environment of a species could be critically altered. It might be due to a climatic change in temperature or rainfall pattern, the introduction of noxious gases like sulphur dioxide from volcanoes, competition from a new species, disappearance of a food source, spread of a disease, and so on. Often extinction results from a combination of causes, even if no one of them on its own would have been fatal. Sometimes the change in the environment is so severe that the whole worldwide community of species is affected, causing a mass extinction.

The rate at which species are becoming extinct at the present time might be seen as simply the next mass extinction marking the start of the next period, often called the Anthropocene to reflect human influences on the environment. However, there are major differences from the past. Conservationists estimate that the world's species are becoming extinct at as much as ten times the rate suffered during previous mass extinctions. This is because the rate of environmental change at present is much more rapid and extensive than previous episodes. In the past, many species had time to evolve to meet the relatively slower changing conditions, or at least to migrate to more suitable areas. Now this is less likely because populations are collapsing too quickly, and because human activities are making more and more of the Earth unavailable for practically any organisms. Before human activity, an increase in global temperature, or a changed pattern of vegetation, would still leave plenty of viable habitats for animals that were able to adapt to them. Today, the environmental deterioration includes a large and growing loss of areas of natural habitat due to intensive agriculture and urbanization, areas in which a balanced, mixed community of species can no longer survive. On top of this pollutants in the atmosphere and oceans also tend to be inimical to all life.

Present threats

Against this background, how are the reptiles faring? So far, very few of the living reptile species ever described are now completely extinct in the wild. Only thirty such are named in the International Union for Conservation of Nature (IUCN) 2017 Red List of species regarded as at risk of extinction. Almost all of these were island species, such as the Seychelles mud turtle that was last seen in 1895, and the Christmas Island whiptailed skink whose last recorded specimen disappeared as recently as 2010. However, a great many more reptiles, around 1,200, are listed in the categories of 'critically endangered', 'endangered', or 'vulnerable', which means that without any positive action they are unlikely to survive for very much longer. Amongst the critically endangered is the Chinese alligator, a fairly typical example. Once occurring throughout most of China, it has been reduced to a tiny population on the southern Yangtze River. Its demise has been caused mainly by the widespread conversion of its wetland habitat into rice paddies, coupled with aggressive persecution by farmers who believe it to be harmful to livestock. The alligators also suffer from eating rats that have been poisoned and, as if all this were not enough, they are often killed for consumption, in the belief that their meat cures cancer.

The giant tortoise species of the Galapagos Islands are another critically endangered iconic example. To take the Santiago tortoise, the numbers say it all. It had a population of around 24,000 in the 19th century, but by 1970 this was down to about 600, representing a 97.5 per cent fall. In this case, the main cause of the decline was the introduction by humans of rats and pigs that devoured the eggs and young tortoises, and of goats and donkeys that outcompeted them for food. The hawksbill turtle is another familiar chelonian on the critically endangered list, its population having declined by over 85 per cent. One of the main threats to this species is posed by the trade in tortoiseshell, its beautiful polished carapace. Until recent voluntary reductions were agreed, hundreds of thousands of turtles were being taken every year, most of them to feed the Japanese market. Other hazards faced by hawksbill turtles are collection of their eggs for food, loss of nesting areas as recreational beaches are developed, and accidental capture in fishing nets. In some places, the development of buildings along turtle nesting beaches has had an unforeseen consequence on the hatchling turtles. Normally they use the lighter sky over the sea as their guide down the beach to the water's edge, but here they are disoriented by the artificial light and head off up the beach instead, to a virtually certain death.

There are certainly a few reptile species that have done well as a result of human activities. House geckos, for example, live in protected, food-rich accommodation, and are very well tolerated by people for their very habit of catching flies. The Florida king snake feeds on rats, and its population has increased with the spread of sugar cane plantations that encourage the rats. It might even be thought that a rise in global temperatures, and the spread of semiarid and arid regions such as in sub-Saharan Africa, would suit those many reptiles that are already adapted for hot, dry conditions. However, the evidence indicates that this is not the case, and that such reptile populations are also falling due to certain side effects of the higher temperatures. One of these is that a raised daily temperature increases the time that a reptile must seek shade to prevent overheating, which in turn reduces the time available for foraging. This is especially critical for females prior to the breeding season, when they need to build up food reserves ready to produce eggs. Another effect can be due to the way temperature controls the sex of the developing offspring. A change may result in an imbalance between males and females.

On balance, we may be sure that the reptile fauna of the world is facing the threat of a considerable reduction in the number of species, although it is impossible to be sure how many will become extinct over the next decades. One estimate is that by the year 2050 over 500 species will have been lost, which is around 5 per cent. By 2080, the figure will have grown to 20 per cent, which is to say about 2,000 species.

As the examples we have just seen of critically vulnerable species show, the causes of the decline and increasing vulnerability to extinction of the reptile fauna are numerous. They are almost entirely consequences of the activities of the inexorable rise in the human population. When, 8,000 years ago, humans first began to practise agriculture and to develop the permanent settlements that proceeded from it, the world's population was about five million. By 1800 it was one billion, today it is seven and a half billion, and by 2050 is estimated to be ten billion. The demand for living space and resources rises proportionately.

Future prospects

The threats to reptiles are of four main kinds: commercial exploitation for food, medicines, and ornament; habitat destruction; global climate change; and pollution. Any comprehensive effort to conserve them needs to address all these. To develop a global strategy, the immediate need is for a great deal more research into the present status of as many reptile species as possible. For far too many of them, we do not even know what the population size is, let alone whether it is declining at a critical rate or not. Also, it is unrealistic to hope that over the coming decades all reptile species can be saved from extinction, and therefore we need to know how we can most effectively focus our efforts. Factors to take into account include how rich in species a particular area is; how likely a region is to be adequately protected from poaching, illegal timber felling, and damaging encroachment of agriculture; and whether a species is particularly unusual or one of numerous similar ones.

By far the most important way to conserve reptiles, along with all other forms of wildlife in their natural habitat, is setting up and regulating various kinds of protected area. To be effective these need to be large enough to sustain viable numbers of all the species present, and to be diverse enough to cover the different habitats of the different animals and plants. Reassuringly, the total area given over to national parks, conservation areas, nature reserves, and other kinds of protected areas has increased steadily and is now a rather impressive 12 per cent or so of the Earth's surface. It should be added, though, that not all are equally well policed.

A development in protecting areas of particular importance is the idea of parks in adjacent nations combining to increase the continuous area and to protect dispersal and migration routes. For example, the Kavango–Zambezi Transfrontier Conservation

Area in southern Africa combines adjacent existing national parks in Namibia, Botswana, and Zambia.

Another important approach to conservation is legislation to control trade in reptiles. The Convention on International Trade in Endangered Species of Wild Fauna and Flora, known as CITES, came into effect in 1975 as a way of regulating and where appropriate prohibiting the commercial exploitation of wildlife. The impact of commercial exploitation on a species is estimated by research, and then the species is placed in one of three appendices depending on how strictly trade should be controlled. At present, eighty-seven reptile species are in Appendix I, which bans all trade in them. Appendices II and III list 810 species that are judged to be in need of strict control to maintain their survival, and they can only be traded under permit. The value of CITES is not only that it attempts to reduce legal and halt illegal trade, but also that it gives government and conservation agencies a clear international legal framework against poaching and smuggling.

From the point of view of particular endangered species, many efforts over the past few years have been devoted to saving them specifically. The unique tuatara, *Sphenodon*, that once existed throughout New Zealand is an excellent example we met in Chapter 3. Almost on the brink of extinction, due mainly to introduced rats eating their eggs, they had become restricted to a few remote offshore islands. But individuals have since been successfully reintroduced to other islands, and to fenced off areas of the mainland from which the rats have first been exterminated. Another case is the artificial breeding of river turtles in Columbia, and the successful reintroduction of the juveniles into the wild.

Most conservation biologists believe that the greatest realistic hope for effective conservation lies in the twin socio-economic factors of education and economics. Wild reptiles are more badly affected than other organisms due to a variety of cultural practices and misunderstandings. A great many reptile species are used in traditional medicine. For example in Morocco, monitor lizards, Mediterranean chameleons, Nile crocodiles, spur-thighed tortoises, and several snakes can be found for sale in the markets for curing all manner of ailments. Few if any of these have any proven therapeutic value compared to modern medicine, yet several are suffering serious decline in their populations as a consequence. Enlightened education of the next generation is the best way to overcome such prejudice.

Crocodile, lizard, and snake skins, and turtle carapaces, are very beautiful materials, but for most of them the supply is readily provided by farming rather than hunting the diminishing numbers of wild specimens, as for example the flourishing crocodile farms in several parts of the world. Similarly, breeding reptiles for the pet market can both be profitable and reduce the stress on the wild populations.

The most optimistic prospect of saving reptiles, along with all the world's wildlife, may lie indirectly in the huge interest in wildlife films, and the growth of ecotourism that follows it. The potential income can be a significant proportion of a nation's GDP, but it depends entirely on developing and sustaining places where wildlife can be satisfactorily and harmlessly experienced by visitors in its true habitat. In some parts of the world, national economic self-interest has already recognized that there is more to be gained from long-term conservation than from short-term exploitation of our ever-diminishing natural world. It is to be hoped that this message spreads in time.

Further reading

Attenborough, D. 2008. *Life in cold blood*. London: BBC Books.

Greene, H. W. 1997. *Snakes: the evolution of mystery in nature*.
Berkeley: California University Press.

Grenard, S. 1991. *Handbook of alligators and crocodiles*. Malabar,
Fla: Krieger Publishing.

Halliday, T. and Adler, K. (eds) 2002. *The new encyclopedia of reptiles
and amphibians*. Oxford: Oxford University Press.

Lillywhite, H. B. 2014. *How snakes work: structure, function and
behavior of the world's snakes*. Oxford: Oxford University Press.

Norman, D. 2017. *Dinosaurs: a very short introduction*, 2nd edition.
Oxford: Oxford University Press.

Orenstein, R. 2012. *Turtles, tortoises and terrapins: a natural history*.
Ontario: Firefly Books.

Pianka, E. R. and Vitt, L. J. 2003. *Lizards: windows to the evolution
of diversity*. Berkeley: California University Press.

Pough, H. P. 2016. *Herpetology*, 4th edition. Sunderland, Mass.:
Sinauer.

Pough, H. P., Janis, C. M., and Heiser, J. B. 2013. *Vertebrate life*,
9th edition (chapters 11–16). Boston: Pearson Education.

Index

Reptiles

Index

Reptiles

SOCIAL MEDIA
Very Short Introduction

Join our community
www.oup.com/vsi

- Join us online at the official Very Short Introductions **Facebook** page.
- Access the thoughts and musings of our authors with our online **blog**.
- Sign up for our monthly **e-newsletter** to receive information on all new titles publishing that month.
- Browse the full range of Very Short Introductions online.
- Read **extracts** from the Introductions for free.
- If you are a teacher or lecturer you can order inspection copies quickly and simply via our website.

DESERTS
A Very Short Introduction
Nick Middleton

Deserts make up a third of the planet's land surface, but if you picture a desert, what comes to mind? A wasteland? A drought? A place devoid of all life forms? Deserts are remarkable places. Typified by drought and extremes of temperature, they can be harsh and hostile; but many deserts are also spectacularly beautiful, and on occasion teem with life. Nick Middleton explores how each desert is unique: through fantastic life forms, extraordinary scenery, and ingenious human adaptations. He demonstrates a desert's immense natural beauty, its rich biodiversity, and uncovers a long history of successful human occupation. This *Very Short Introduction* tells you everything you ever wanted to know about these extraordinary places and captures their importance in the working of our planet.

HIV/AIDS
A Very Short Introduction
Alan Whiteside

HIV/AIDS is without doubt the worst epidemic to hit humankind since the Black Death. The first case was identified in 1981; by 2004 it was estimated that about 40 million people were living with the disease, and about 20 million had died. The news is not all bleak though. There have been unprecedented breakthroughs in understanding diseases and developing drugs. Because the disease is so closely linked to sexual activity and drug use, the need to understand and change behaviour has caused us to reassess what it means to be human and how we should operate in the globalising world. This *Very Short Introduction* provides an introduction to the disease, tackling the science, the international and local politics, the fascinating demographics, and the devastating consequences of the disease, and explores how we have — and must — respond.

'It won't make you an expert. But you'll know what you're talking about and you'll have a better idea of all the work we still have to do to wrestle this monster to the ground.'

Aids-free world website.

www.oup.com/vsi

GALAXIES
A Very Short Introduction
John Gribbin

Galaxies are the building blocks of the Universe: standing like islands in space, each is made up of many hundreds of millions of stars in which the chemical elements are made, around which planets form, and where on at least one of those planets intelligent life has emerged. In this *Very Short Introduction*, renowned science writer John Gribbin describes the extraordinary things that astronomers are learning about galaxies, and explains how this can shed light on the origins and structure of the Universe.

www.oup.com/vsi

GLOBALIZATION
A Very Short Introduction
Manfred Steger

'Globalization' has become one of the defining buzzwords of our time - a term that describes a variety of accelerating economic, political, cultural, ideological, and environmental processes that are rapidly altering our experience of the world. It is by its nature a dynamic topic - and this *Very Short Introduction* has been fully updated for 2009, to include developments in global politics, the impact of terrorism, and environmental issues. Presenting globalization in accessible language as a multifaceted process encompassing global, regional, and local aspects of social life, Manfred B. Steger looks at its causes and effects, examines whether it is a new phenomenon, and explores the question of whether, ultimately, globalization is a good or a bad thing.

www.oup.com/vsi

SLEEP
A Very Short Introduction
Russell G. Foster & Steven W. Lockley

Why do we need sleep? What happens when we don't get enough? From the biology and psychology of sleep and the history of sleep in science, art, and literature; to the impact of a 24/7 society and the role of society in causing sleep disruption, this *Very Short Introduction* addresses the biological and psychological aspects of sleep, providing a basic understanding of what sleep is and how it is measured, looking at sleep through the human lifespan and the causes and consequences of major sleep disorders. Russell G. Foster and Steven W. Lockley go on to consider the impact of modern society, examining the relationship between sleep and work hours, and the impact of our modern lifestyle.

www.oup.com/vsi

THE HISTORY OF
MEDICINE
A Very Short Introduction
William Bynum

Against the backdrop of unprecedented concern for the future of
health care, this Very Short Introduction surveys the history of
medicine from classical times to the present. Focussing on the
key turning points in the history of Western medicine, such as the
advent of hospitals and the rise of experimental medicine, Bill
Bynum offers insights into medicine's past, while at the same time
engaging with contemporary issues, discoveries, and
controversies.